T0191771

Progress in Mathematics
Volume 311

More information about this series at http://www.springer.com/series/4848

Vincent Franjou • Antoine Touzé
Editors

Lectures on Functor Homology

Editors
Vincent Franjou
Laboratoire de Mathématiques Jean Leray
Université de Nantes
Nantes, France

Antoine Touzé
Laboratoire Paul Painlevé
Université Lille 1
Villeneuve d'Ascq, France

ISSN 0743-1643 ISSN 2296-505X (electronic)
Progress in Mathematics
ISBN 978-3-319-79333-7 ISBN 978-3-319-21305-7 (eBook)
DOI 10.1007/978-3-319-21305-7

Mathematics Subject Classification (2010): 14L, 18E, 18G, 19D55, 20G10, 20J, 55U99, 55Q

Springer Cham Heidelberg New York Dordrecht London

Softcover re-print of the Hardcover 1st edition 2015

Printed on acid-free paper

Springer International Publishing AG Switzerland is part of Springer Science+Business Media (www.birkhauser-science.com)

Contents

Progress in Mathematics, Vol. 311, 1–6

Introduction

Vincent Franjou and Antoine Touzé

This book is an account of series of lectures held at the conference on functor homology, which took place in Nantes in April, 2012. Functor homology is shorthand for homological algebra in functor categories, a topic that potentially covers a large field of interests. The series presented three different fields where functor homology has recently played a significant role. For each of these applications, the functorial viewpoint provides both insights and new methods for tackling difficult mathematical problems.

The central role is played by the notion of polynomial functor and variants of this notion. The notion of polynomial functors has two quite independent origines. The first one from the work of Schur [Sch01] in representation theory. The second one from the work of Eilenberg and Mac Lane [EML54] in algebraic topology. For precise definitions, the reader can wait and read the presentation in each lecture. For this introduction, let us insist on the idea that polynomial functors are tame functors, good companions like the symmetric powers S^d or the exterior powers Λ^d, considered as endofunctors of the category of modules over a commutative ring R.

In the first series of lectures by Aurélien Djament, polynomial functors appear in connection with the (co)homology of classical groups and their stabilization. To be specific, let us take the family of symplectic groups $Sp_n(R)$. A group $Sp_n(R)$ acts naturally on R^{2n}, defining its standard representation. The dth symmetric power of this representation, $S^d(R^{2n})$, inherits an action of $Sp_n(R)$ – mind it, this is a consequence of the fact that the symmetric power is a functor. Consider the (co)homology of the discrete group $Sp_n(R)$ with coefficients in this representation; this is denoted by: $\mathrm{H}(Sp_n(R), S^d(R^{2n}))$. Djament's theorem tells us that this homological invariant, if n is large enough, can be computed using only the homology with trivial coefficients $\mathrm{H}(Sp_n(R))$, and homological computations in the category of functors $\mathcal{F}_R$. The fact that the symmetric power functor S^d is polynomial is crucial in the proof of this result, and the result is indeed valid when S^d is replaced by any other polynomial functor. This provides much more than a conceptual expression of generic homology: such a bridge between stable homology of classical groups and functor homology is effective to perform computations. Indeed, several surprisingly powerful tools are available to do homological algebra in $\mathcal{F}_R$, as was explained in the collected lectures [FFPS03]. Djament's results extends our understanding of group cohomology by functor cohomology, as initiated by Suslin [FFSS99, Appendix] and Scorichenko [Sco00].

Wilberd van der Kallen's series of lectures study the cohomology of algebraic groups. This cohomology is sometimes called rational cohomology, or algebraic cohomology. The cohomology of an algebraic group G is quite different [Tou, Ex. 2.57] from the cohomology of G considered as a discrete group, that we have encountered in the previous paragraph. In particular its coefficients are rational (or algebraic) representations of the group G. However, the two cohomologies, the cohomology of the discrete group and the algebraic cohomology, are related. This is the topic of the celebrated result of Cline, Parshall, Scott, and van der Kallen [CPSvdK77] when R is a finite field. Friedlander and Suslin [FS97] designed a category of functors to study the cohomology of algebraic groups, and of $GL_n(R)$ in particular. It is the category of strict polynomial functors $\mathcal{P}_R$. The category $\mathcal{P}_R$ is an algebraic analogue of the category of ordinary functors $\mathcal{F}_R$, and the relation between the cohomology of classical algebraic groups on one hand, and the cohomology of their discrete counterpart on the other hand, can be seen and measured at the level of the corresponding functor categories [FFSS99]. The relevance of functor cohomology calculations to algebraic cohomology has been a major step. It allowed Friedlander and Suslin to prove finite generation of the cohomology of finite group schemes [FS97] and thus initiate a theory of support varieties [SFB97]. It eventually lead to the solution [TvdK10] of the general cohomological finite generation problem, the generalization of Hilbert's fourteenth problem presented in van der Kallen's series.

Roman Mikhailov's series of lectures also show polynomial functors and homological algebra in action, although the topic and its treatment might seem quite different from the two previous ones. The aim here is to compute topological invariants: homotopy, homology of topological spaces, through derived functors of non-additive functors [DP61]. Of course, the non-additive functors we have in mind here are polynomial functors like S^d and Λ^d.

An emblematic example of functorial techniques is Simson and Tyc's use [ST70] of the (exact) Koszul complex:

$$0 \to \Lambda^d \to \cdots \Lambda^{d-i} \otimes S^i \to \Lambda^{d-i-1} \otimes S^{i+1} \to \cdots S^d \to 0$$

together with the fact that the stable derived functors $L_*^{st}(F \otimes G)$ of a tensor product of reduced functors vanish [DP61, Korollar 5.20]. They prove an isomorphism, of degree $d-1$ (or décalage)

$$L_*^{st}\Lambda^d(-) \simeq L_*^{st}S^d(-) \ .$$

Indeed, to prove the décalage, slice the Koszul complex into successive short exact sequences. Then, on each short exact sequence, use the long exact sequence for the stable derived functors L_*^{st}. Simson's argument was successfully imported (or rediscovered) in the world of functors. Indeed, by the fundamental result of Pirashvili [Pir88] the cohomology groups, in $\mathcal{F}_R$, of a tensor product $\mathrm{Ext}^*(S^1, F \otimes G)$

similarly vanish. The argument was extended to compute [FLS94] the algebra $Ext^*_{\mathcal{F}_R}(S^1, S^1)$ (and many other cohomology groups in $\mathcal{F}_R$ as well). Scorichenko's vanishing theorem, as presented in A. Djament's third lecture, is an advanced form of this sort of vanishing result.

Derived functors of non-additive functors are also directly related to functor cohomology in the category $\mathcal{P}_R$ of *strict* polynomial functors. For example, a correspondence recently found [Tou13b] yields natural isomorphisms

$$L_iF(R,1) \simeq \mathrm{Ext}^{d-i}_{\mathcal{P}_R}(\Lambda^d, F)\ , \qquad L_iF(R,2) \simeq \mathrm{Ext}^{2d-i}_{\mathcal{P}_R}(S^d, F)\ ,$$

for all homogeneous strict polynomial functors F of degree d.

Roman Mikhailov's series of lectures appeals to modern theory and makes better use of naturality, to reach calculations, that classical theory could not reach. For example, natural transformations and extension groups between functors are made explicit through a new combinatorial description of polynomial functors [BDFP01]. Such combinatorial descriptions of polynomial functors are now available in a much broader context [HPV]. They rely on Pirashvili's Dold–Kan type theorem [Pir00], that was unavailable to Eilenberg and Mac Lane in their original definition of polynomial functors [EML54]. Modern theory thus feeds back the classical theory.

The links between functor homology and the three fields mentioned above reveal common structure. However, each of these fields requires specific adaptations, and therefore they enrich the functorial viewpoint. These applications of functor homology do not constitute an exhaustive list. Other applications already exists, for example to the Steenrod algebra [FFPS03], and hopefully new applications will be found in the future. By gathering these lectures together, we hope to give the reader a feeling of the functorial viewpoint, and new insight to apply on his favourite problems.

We now focus on the specific content of the different parts of this book.

Aurélien Djament's lectures explains correspondences between functor homology, and stable homology (with coefficients) of families of discrete groups. Djament's first lecture presents a general setup [DV10, Dja12] designed to simply lead to explicit and highly nontrivial results. The lectures thus give a smooth introduction to the field, albeit providing all the technical details, and they encompass various fundamental results of Betley, Suslin, and of Scorichenko. This is already illustrated by Lecture 2. Scorichenko's cancellation theorem [Sco00] is generalized and strengthened in Lecture 3. It is presented in a way that makes this unpublished result available for general use. Lecture 4 contains the main result. The text also presents motivations, and it includes in the final lecture sample applications and calculations. On the way, the text surveys other related topics, such as stable homology of groups with trivial coefficients. The density of the lectures and the clarity of exposition meet the ambition of a reference paper on the subject.

Wilberd van der Kallen's lectures deal with the proof [TvdK10] of the cohomological finite generation conjecture for reductive algebraic groups. The lectures start with a historical introduction. They proceed by introducing and explaining many mathematical objects which play a role in the conjecture's proof: representations of group schemes and cohomology, with an emphasis on the General Linear Group; and Friedlander's and Suslin's strict polynomial functors. The last lecture presents a key ingredient in the proof of finite generation, that is the construction of certain "universal cohomology classes" using functor category tools. The lectures thus provide the non-expert reader with the background she needs when reading the original proof.

The core of the lectures, however, deals with a related but slightly different topic, namely a formality conjecture of Chałupnik. The formality conjecture is a strong form of a conjecture by Touzé on the effect of Frobenius twists on extension groups, in Friedlander's and Suslin's category of strict polynomial functors. This is strongly connected to cohomological finite generation, since this description leads [Tou13a] to a second generation, more conceptual, proof of the existence of the universal classes, which come as a key ingredient to prove the cohomological finite generation. The penultimate lecture by van der Kallen presents a proof of the formality conjecture, and it clarifies several of its more subtle points.

Roman Mikhailov's lectures start with a presentation of polynomial functors between abelian groups. Mikhailov takes seriously the combinatorics of polynomial functors in [BDFP01]. One of the innovative points of his lectures is the use of this point of view to perform computations. He applies this technique to problems coming from algebraic topology: the functorial description of low degree homology of abelian groups in Section 3, or of low stable homotopy of Eilenberg–Mac Lane spaces in Section 4, and related algebraic problems about derived functors of non-additive functors in Section 5. The last section gives examples of the use of naturality to compute spectral sequences (differentials and extension problems).

The book ends with a short course in homological algebra, by Antoine Touzé. The first part of the course is a presentation of derived functors. This concept allows a unified presentation of the many different objects of study of the book. Emphasis is then put on the analogies, similarities, and also differences between the various (co)homological notions that appear in the three main series of lectures. The second part of the course is a short introduction to spectral sequences, from the user's viewpoint. Spectral sequences are especially useful when coming to explicit computations, and they prove so in every series of the book. Little originality is claimed here. The purpose is to ease access to the book for everyone, and to allow this book to be used as a first course in homological algebra for the absolute beginner. We believe that meeting such objectives largely contributed to the success of the meeting in Nantes.

References

[BDFP01] H.-J. Baues, W. Dreckmann, V. Franjou, and T. Pirashvili. Foncteurs polynomiaux et foncteurs de Mackey non linéaires. *Bull. Soc. Math. France*, 129(2):237–257, 2001.

[CPSvdK77] E. Cline, B. Parshall, L. Scott, and W. van der Kallen. Rational and generic cohomology. *Invent. Math.*, 39(2):143–163, 1977.

[Dja12] A. Djament. Sur l'homologie des groupes unitaires à coefficients polynomiaux. *J. K-Theory*, 10(1):87–139, 2012.

[DP61] A. Dold and D. Puppe. Homologie nicht-additiver Funktoren. Anwendungen. *Ann. Inst. Fourier Grenoble*, 11:201–312, 1961.

[DV10] A. Djament and C. Vespa. Sur l'homologie des groupes orthogonaux et symplectiques à coefficients tordus. *Ann. Sci. Éc. Norm. Supér.* (4), 43(3):395–459, 2010.

[EML54] S. Eilenberg and S. Mac Lane. On the groups $H(\Pi, n)$. II. Methods of computation. *Ann. of Math.* (2), 60:49–139, 1954.

[FFPS03] V. Franjou, E.M. Friedlander, T. Pirashvili, and L. Schwartz. *Rational representations, the Steenrod algebra and functor homology*, volume 16 of *Panoramas et Synthèses [Panoramas and Syntheses]*. Société Mathématique de France, Paris, 2003.

[FFSS99] V. Franjou, E.M. Friedlander, A. Scorichenko, and A. Suslin. General linear and functor cohomology over finite fields. *Ann. of Math.* (2), 150(2):663–728, 1999.

[FLS94] V. Franjou, J. Lannes, and L. Schwartz. Autour de la cohomologie de Mac Lane des corps finis. *Invent. Math.*, 115(3):513–538, 1994.

[FS97] E.M. Friedlander and A. Suslin. Cohomology of finite group schemes over a field. *Invent. Math.*, 127(2):209–270, 1997.

[HPV] M. Hartl, T. Pirashvili, and C. Vespa. Polynomial functors from algebras over a set-operad and non-linear mackey functors. *Int. Math. Res. Not.*, (to appear).

[Pir88] T. Pirashvili. Higher additivizations. *Trudy Tbiliss. Mat. Inst. Razmadze Akad. Nauk Gruzin. SSR*, 91:44–54, 1988.

[Pir00] T. Pirashvili. Dold–Kan type theorem for Γ-groups. *Math. Ann.*, 318(2):277–298, 2000.

[Sch01] I. Schur. *Thesis* (1901), volume Gesammelte Abhandlungen. Band I. Springer-Verlag, Berlin-New York, 1973. Herausgegeben von Alfred Brauer und Hans Rohrbach.

[Sco00] A. Scorichenko. *Stable K-theory and functor homology over a ring*. PhD thesis, Evanston, 2000.

[SFB97] A. Suslin, E.M. Friedlander, and C.P. Bendel. Infinitesimal 1-parameter subgroups and cohomology. *J. Amer. Math. Soc.*, 10(3):693–728, 1997.

[ST70] D. Simson and A. Tyc. On stable derived functors. II. *Bull. Acad. Polon. Sci. Sér. Sci. Math. Astronom. Phys.*, 18:635–639, 1970.

[Tou] A. Touzé. Prerequisites of homological algebra. This volume.

[Tou13a] A. Touzé. A construction of the universal classes for algebraic groups with the twisting spectral sequence. *Transform. Groups*, 18(2):539–556, 2013.

[Tou13b] A. Touzé. Ringel duality and derivatives of non-additive functors. *J. Pure Appl. Algebra*, 217(9):1642–1673, 2013.

[TvdK10] A. Touzé and W. van der Kallen. Bifunctor cohomology and cohomological finite generation for reductive groups. *Duke Math. J.*, 151(2):251–278, 2010.

Vincent Franjou
Université de Nantes
Laboratoire de Mathématiques Jean Leray (UMR 6629)
Faculté des Sciences
2, rue de la Houssinière
BP 92208
F-44322 Nantes cedex 3, France
e-mail: `Vincent.Franjou@univ-nantes.fr`

Antoine Touzé
LAGA, CNRS (UMR 7539) Université Paris 13
99, Av. J.-B. Clément
F-93430 Villetaneuse, France
e-mail: `Touze@math.univ-paris13.fr`

Progress in Mathematics, Vol. 311, 7–39

Homologie stable des groupes à coefficients polynomiaux

Aurélien Djament

Abstract. This series of lectures deals with stable homology of (discrete) groups with coefficients twisted by a suitable functor. By *stable* homology we mean the colimit of the homology of a nice sequence of groups, such as the symmetric groups or as the general linear groups over a fixed ring. The lectures' aim is to express stable homology with twisted coefficients using stable homology with untwisted coefficients and functor homology.

Mathematics Subject Classification (2010). 18A25, 20J06, 18G15, 18A40.

Keywords. Group homology, functor categories, polynomial functors, cross effects, hermitian spaces, derived Kan extension.

Remarque. Les parties écrites en caractères linéaux peuvent être omises en première lecture.

Les cinq sections (correspondant aux cinq cours) sont conçues pour pouvoir être abordées de manière assez largement indépendante.

1. Introduction

Abstract. *The introductory lecture sets the stage (the groups are the automorphism groups of sums of copies of a given object in a suitable symmetric monoidal category), and it gives theoretic results and computational or qualitative consequences of the good features of functor homology. It also gives a quick overview of a few related topics (homological stability, algebraic K-theory...).*

Ce mini-cours traite d'homologie des groupes *discrets*, essentiellement à coefficients tordus (i.e., avec une action non triviale du groupe sur les coefficients). On s'intéresse surtout à des phénomènes *stables*, ou *génériques*, c'est-à-dire qui ne

L'auteur sait gré à Christine Vespa, Antoine Touzé et Vincent Franjou d'échanges utiles à l'amélioration des premières versions de ce texte.

concernent pas un groupe en particulier (dont on saura souvent dire très peu du point de vue homologique), mais une famille de groupes.

Précisons un peu. Certaines tours de groupes, i.e., suites $(G_n)_{n\in\mathbb{N}}$ munies de morphismes de groupes $G_n \to G_{n+1}$, sont particulièrement intéressantes du point de vue de l'homologie. Pour nous, il s'agira essentiellement des suites suivantes :

1. la suite des groupes symétriques Σ_n (permutations de l'ensemble $\{1,\ldots,n\}$), où Σ_n est plongé dans Σ_{n+1} comme le sous-groupe des permutations laissant $n+1$ invariant ;
2. la suite des groupes linéaires $GL_n(A)$ sur un anneau A, où $GL_n(A)$ est plongé dans $GL_{n+1}(A)$ par $M \mapsto \begin{pmatrix} M & 0 \\ 0 & 1 \end{pmatrix}$.
3. la suite des groupes orthogonaux $O_{n,n}(A)$ ou des groupes symplectiques $Sp_{2n}(A)$, où A est un anneau commutatif, où les plongements sont donnés comme précédemment – on considère ici par convention les formes quadratique et symplectique dont les matrices sont constituées de n blocs diagonaux égaux à $\begin{pmatrix} 0 & 1 \\ 1 & 0 \end{pmatrix}$ et $\begin{pmatrix} 0 & 1 \\ -1 & 0 \end{pmatrix}$ respectivement (cet exemple est susceptible de variations évidentes).

La suite de groupes d'homologie

$$H_*(G_0) \to H_*(G_1) \to \cdots \to H_*(G_n) \to H_*(G_{n+1}) \to \cdots$$

(volontairement, on ne précise pas les coefficients à ce stade) donne lieu à plusieurs questions fondamentales :

1. sait-on la calculer complètement ? (La réponse n'est qu'exceptionnellement positive, même lorsque les coefficients sont constants, néanmoins cela peut arriver dans quelques cas particulièrement favorables.)
2. Sait-on calculer sa colimite, appelée *homologie stable* de la suite de groupes ? (C'est encore, souvent, une question très difficile, mais on sait dire beaucoup plus de choses que pour la question précédente.)
3. La suite se stabilise-t-elle (c'est-à-dire que, pour tout $d \in \mathbb{N}$, il existe N tel que $H_d(G_n) \to H_d(G_{n+1})$ soit un isomorphisme pour $n \geq N$), et si oui avec quelles bornes ? (L'un des principaux objectifs est, en l'absence de réponse positive à la première question, de tirer des renseignements sur $H_*(G_n)$, à n fixé, de l'homologie stable.)

Hormis dans cette première séance introductive, nous n'aborderons que la deuxième question, *dans le cas de coefficients tordus favorables*, en supposant connue l'homologie stable à coefficients constants : c'est ce pour quoi l'homologie des foncteurs a montré son efficacité. Pour l'instant, rien de convaincant n'a été réalisé à l'aide de l'homologie des foncteurs pour aborder l'homologie à coefficients constants de groupes discrets ; toutefois, les raisons conceptuelles de la dichotomie observée entre coefficients constants et tordus (polynomiaux en un sens adéquat, en général) demeurent obscures.

Remarque 1.1. Nous ne parlerons le plus souvent que d'homologie, qui possède la propriété commode de commuter aux colimites filtrantes (ainsi, on peut voir l'homologie stable de notre tour de groupes comme l'homologie de sa colimite) ; néanmoins, des résultats tout à fait analogues valent en cohomologie.

L'un des intérêts de l'homologie stable est de *structurer* en général davantage la situation que l'homologie individuelle d'un des groupes. Ainsi, dans tous les cas susmentionnés, au moins lorsque les coefficients sont pris sur un corps commutatif (avec action triviale), on dispose d'une structure d'algèbre de Hopf graduée connexe commutative et cocommutative. Ce type d'observation peut s'avérer crucial pour aborder les problèmes posés (par exemple, le calcul de l'homologie des groupes symétriques présenté dans [AM04] utilise les résultats de structure des algèbres de Hopf sur $\mathbb{F}_p$).

Remarque 1.2. Cette situation rappelle d'une certaine manière celle de l'homotopie stable par rapport à l'homotopie instable : on dispose de résultats de stabilisation de la colimite qui définit les groupes d'homotopie stable à partir des groupes d'homotopie ordinaires pour des espaces possédant de bonnes propriétés de finitude (théorème de Freudenthal) ; la catégorie homotopique des spectres possède une structure plus riche et maniable (c'est une catégorie triangulée) que la catégorie homotopique des espaces topologiques pointés, qui n'est même pas additive.

1.1. Un cadre raisonnable : homologie de groupes d'automorphismes dans une catégorie monoïdale symétrique

(On renvoie à [ML98] pour les définitions et propriétés fondamentales des catégories monoïdales. Le cadre présenté ici, comme une partie significative des résultats donnés dans les exposés suivants, provient de [DV10].)

Soit $(\mathcal{C}, \oplus, 0)$ une catégorie monoïdale symétrique (bien qu'on note $\oplus$ le foncteur donnant la structure monoïdale, on ne suppose pas qu'il s'agit d'une somme catégorique). On suppose que l'unité 0 est objet initial (mais pas nécessairement nul) de $\mathcal{C}$ [1]. On se donne par ailleurs un objet X de $\mathcal{C}$. Les groupes auxquels on s'intéresse sont les groupes d'automorphismes de la « somme » (au sens de $\oplus$) de copies de X :

$$G_n := \mathrm{Aut}_{\mathcal{C}}(X^{\oplus n}).$$

On rappelle que l'on peut sans restriction supposer que $\oplus$ est strictement associatif ; en revanche, malgré la symétrie, il faut toujours prendre garde à l'ordre des facteurs. De fait, la symétrie procure un morphisme de groupes $\Sigma_n \to G_n$ (action par permutation des facteurs de la somme).

On fait des G_n une tour de la façon suivante : à un automorphisme u de $X^{\oplus n}$ on associe l'automorphisme $u \oplus Id_X$ de $X^{\oplus n+1}$, ce qui définit un morphisme de groupes $G_n \to G_{n+1}$. Noter que l'on aurait pu choisir d'associer $Id_X \oplus u$, par

1. Rappelons la signification du fait que 0 est objet initial : pour tout objet c de $\mathcal{C}$, il existe un et un seul morphisme de source 0 et de but c. Exiger que 0 soit objet final reviendrait à demander qu'il existe un et un seul morphisme de source c et de but 0, pour tout objet c de $\mathcal{C}$. Un objet nul est un objet à la fois initial et final.

exemple, à u plutôt que $u \oplus Id_X$, ce qui fournit un morphisme de groupes *différent.* Néanmoins, ces deux choix sont *conjugués* sous l'action de la transposition $(1 \quad n+1)$ sur $X^{\oplus n+1}$, de sorte qu'ils induisent le *même* morphisme $H_*(G_n) \to H_*(G_{n+1})$ en homologie. Cette observation sera utilisée abondamment.

Exemple 1.3.

1. La catégorie des ensembles finis munie de la somme catégorique (réunion disjointe), avec A ensemble à un élément, donne les groupes symétriques. Plusieurs variantes fournissent la même tour de groupes : on peut considérer la sous-catégorie monoïdale avec les mêmes objets et les injections comme morphismes, ou des catégories d'ensembles finis pointés (avec la somme pointée comme structure monoïdale et en prenant pour X un ensemble pointé à deux éléments).
2. Si A est un anneau, la catégorie $\mathbf{P}(A)$ des A-modules à gauche projectifs de type fini munie de la somme directe donne, avec $X = A$, la tour des $GL_n(A)$. Nous reviendrons plus tard sur d'autres variantes de catégories de A-modules projectifs éventuellement plus adaptées.
3. Soient A est un anneau muni d'une anti-involution, $\epsilon \in \{-1, 1\}$, et $\mathbf{H}_\epsilon(A)$ la catégorie des A-modules à gauche projectifs de type fini munis d'une forme ϵ-hermitienne. Explicitement, si l'on note D l'endofoncteur contravariant $\mathrm{Hom}_A(-, A)$ des A-modules à gauche projectifs de type fini (on utilise l'anti-involution de A pour convertir l'action naturelle de A à droite sur $\mathrm{Hom}_A(M, A)$, où M est un A-module à gauche, en une action à gauche), D possède une propriété d'auto-adjonction : $\mathrm{Hom}_A(M, DN) \simeq \mathrm{Hom}_A(N, DM)$, isomorphisme noté par une barre, qui est une involution pour $M = N$, une forme ϵ-hermitienne sur M est un élément du groupe abélien $\mathrm{S}^2_\epsilon(M)$ quotient de $\mathrm{Hom}_A(M, DM)$ par l'image de l'endomorphisme $x \mapsto \bar{x} - \epsilon x$. (Pour A commutatif avec involution triviale, c'est la notion usuelle d'espace quadratique si $\epsilon = 1$ et symplectique si $\epsilon = -1$.) La structure monoïdale est la somme directe orthogonale ; on prend pour X le A-module A^2 (auquel il faut penser comme $A \oplus DA$) muni de la forme donnée par la classe de la matrice $\begin{pmatrix} 0 & 1 \\ 0 & 0 \end{pmatrix}$. On obtient ainsi, selon que ϵ vaut 1 ou -1, la tour des groupes unitaires (hyperboliques) ou des groupes symplectiques. Noter qu'on pourrait se restreindre à la sous-catégorie des espaces hermitiens non dégénérés (nous reviendrons également en détail sur tout cela ultérieurement).
4. La catégorie des groupes libres de type fini, avec la somme catégorique (produit libre), s'insère également dans ce formalisme.

Remarque 1.4. Dans ces situations, il est naturel de se demander ce qui se passe lorsqu'on remplace les groupes en question par des sous-groupes remarquables comme les groupes alternés dans les groupes symétriques ou les groupes linéaires spéciaux dans les groupes linéaires (sur un anneau commutatif). Ils s'avèrent moins commodes à manier puisqu'ils ne contiennent pas les groupes symétriques, mais peuvent se traiter de façon analogue « à la main » , en utilisant par exemple l'inclusion $GL_n(A) \hookrightarrow SL_{n+1}(A)$ donnée par $M \mapsto \begin{pmatrix} M & 0 \\ 0 & (\det M)^{-1} \end{pmatrix}$.

Montrons comment tordre les coefficients à l'aide d'un foncteur $F : \mathcal{C} \to \mathbf{Ab}$. Utilisant que 0 est objet initial de $\mathcal{C}$, on dispose dans $\mathcal{C}$ de morphismes

$$X^{\oplus n} = X^{\oplus n} \oplus 0 \xrightarrow{Id \oplus (0 \to X)} X^{\oplus n} \oplus X = X^{\oplus n+1}$$

(là encore on pourrait choisir d'inclure d'autres facteurs que les n premiers, mais il faut que le choix soit compatible avec celui effectué pour les morphismes $G_n \to G_{n+1}$), d'où des morphismes $F(X^{\oplus n}) \to F(X^{\oplus n+1})$ compatibles aux actions tautologiques de G_n et G_{n+1} et au morphisme $G_n \to G_{n+1}$. On dispose donc de morphismes de groupes abéliens gradués $H_*(G_n; F(X^{\oplus n})) \to H_*(G_{n+1}; F(X^{\oplus n+1}))$ naturels en F ; la colimite de cette suite de morphismes est appelée *homologie stable de la tour* (G_n) *(ou de* $\mathcal{C}$*) à coefficients dans* F. Comme homologie et colimites filtrantes commutent, cette homologie stable s'identifie canoniquement à $H_*(G_\infty; F_\infty)$, où F_∞ est la colimite des $F(X^{\oplus n})$ et G_∞ celle des G_n.

Le but de ce cours est de montrer comment calculer, dans les exemples fondamentaux précédents, l'homologie stable à partir de $H_*(G_\infty)$ (homologie à coefficients dans $\mathbb{Z}$ de G_∞), pour F suffisamment raisonnable (en particulier *polynomial* – notion sur laquelle nous reviendrons), en utilisant l'homologie $H_*(\mathcal{C}; F)$ de la catégorie $\mathcal{C}$ à coefficients dans F.

Homologie (de Hochschild) d'un (bi)foncteur. Si $\mathcal{C}$ est une petite catégorie (ou une catégorie essentiellement petite) et $\Bbbk$ un anneau commutatif fixé, la catégorie $\mathcal{C}$-**Mod** des foncteurs de $\mathcal{C}$ vers les $\Bbbk$-modules est une catégorie abélienne qui se comporte comme une catégorie de modules (elle a assez d'objets injectifs et projectifs, possèdent des limites et colimites, les colimites filtrantes sont exactes. . .) : on peut y faire de l'algèbre homologique de façon analogue. On dispose notamment de l'homologie d'un foncteur $F \in \mathrm{Ob}\, \mathcal{C}\text{-}\mathbf{Mod}$, notée $H_*(\mathcal{C}; F)$: $H_0(\mathcal{C}; F)$ est simplement la colimite de F, et l'homologie de degré supérieur s'obtient en dérivant à gauche (la colimite est un foncteur exact à droite $\mathcal{C}\text{-}\mathbf{Mod} \to \mathbf{Mod}_\Bbbk$). On dispose également, pour $F \in \mathrm{Ob}\, \mathcal{C}\text{-}\mathbf{Mod}$ et $G \in \mathrm{Ob}\, \mathbf{Mod}\text{-}\mathcal{C} := \mathcal{C}^{\mathrm{op}}\text{-}\mathbf{Mod}$ de groupes de torsion $\mathrm{Tor}_*^{\mathcal{C}}(G, F)$, qui dérivent le produit tensoriel au-dessus de $\mathcal{C}$. On dispose enfin, si B est un bifoncteur sur $\mathcal{C}$, c'est-à-dire un objet de $(\mathcal{C}^{\mathrm{op}} \times \mathcal{C})$-**Mod** (on prendra garde que, selon les sources, c'est parfois la première variable, parfois la seconde qui est contravariante !), de l'homologie de Hochschild de $\mathcal{C}$ à coefficients dans B, notée $HH_0(\mathcal{C}; B)$. En degré 0, c'est la *cofin* de B (cf. [ML98]), en degré supérieur on dérive à gauche. Si F est un foncteur covariant sur $\mathcal{C}$ et G un foncteur contravariant, sur le bifoncteur produit tensoriel extérieur $G \boxtimes F : (A, B) \mapsto G(A) \otimes F(B)$ (les produits tensoriels de base non spécifiée sont pris sur $\Bbbk$), on dispose d'un isomorphisme canonique $HH_0(\mathcal{C}; G \boxtimes F) \simeq G \otimes_{\mathcal{C}} F$ qui s'étend, lorsque F ou G prennent des valeurs plates sur $\Bbbk$, en un isomorphisme naturel gradué $HH_*(\mathcal{C}; G \boxtimes F) \simeq \mathrm{Tor}_*^{\mathcal{C}}(G, F)$. En particulier, $HH_*(\mathcal{C}; F) \simeq \mathrm{Tor}_*^{C}(\Bbbk, F) \simeq H_*(\mathcal{C}; F)$, où $\Bbbk$ désigne le foncteur constant en $\Bbbk$ et F est vu, dans le membre de gauche, comme bifoncteur ne dépendant pas de la variable contravariante. On dispose par ailleurs, pour tout bifoncteur B, d'un isomorphisme naturel $HH_*(\mathcal{C}; B) \simeq \mathrm{Tor}_*^{\mathcal{C}^{\mathrm{op}} \times \mathcal{C}}(\Bbbk[\mathrm{Hom}_{\mathcal{C}^{\mathrm{op}}}], B)$.

1.2. Aperçu des résultats principaux présentés dans ce mini-cours

Si A est un anneau, on note $\mathbf{P}(A)$ la catégorie des A-modules à gauche projectifs de type fini.

Théorème 1.5 (Betley [Bet89]). *Soient A un anneau commutatif et F un foncteur* **polynomial**[2] *de $\mathbf{P}(A)$-**Mod** tel que $F(0) = 0$.*

Alors $H_*(GL_\infty(A); F_\infty) = 0$.

Théorème 1.6 (Betley [Bet99], Suslin [FFSS99] (appendice)). *Soient k un corps fini et B un bifoncteur polynomial sur $\mathbf{P}(k)$; on suppose que l'anneau des coefficients est $\Bbbk = k$. Alors il existe un isomorphisme naturel*

$$H_*(GL_\infty(k); B_\infty) \simeq HH_*(\mathbf{P}(k); B)$$

de k-espaces vectoriels gradués.

Théorème 1.7 (Scorichenko [Sco00]). *Soient A un anneau et B un bifoncteur polynomial sur $\mathbf{P}(A)$ (i.e., un objet de $(\mathbf{P}(A)^{\mathrm{op}} \times \mathbf{P}(A))$-**Mod**). Il existe une suite spectrale naturelle*

$$E^2_{p,q} = H_p(GL_\infty(A); HH_q(\mathbf{P}(A); B)) \Rightarrow H_{p+q}(GL_\infty(A); B_\infty)$$

(où l'action du groupe linéaire est triviale), qui s'effondre à la deuxième page si $\Bbbk$ est un anneau principal.

(Ce résultat est en général énoncé en termes de K-théorie stable, mais nous n'en aurons nul usage, et c'est de fait la propriété de suite spectrale qui importe toujours dans les utilisations. Par ailleurs, l'article [DV10] explique comment construire la suite spectrale sans utiliser la K-théorie stable.)

Théorème 1.8 (Djament–Vespa [DV10]). *Soient k un corps fini de caractéristique différente de 2*[3] *et F un foncteur polynomial de $\mathbf{P}(k)$ vers les k-espaces vectoriels. Il existe un isomorphisme naturel*

$$H_*(O_\infty(k); F_\infty) \simeq \mathrm{Tor}_*^{\mathbf{P}(k)}(k[S^2_k]^\vee, F)$$

où S^2_k désigne la deuxième puissance symétrique sur k et l'exposant $\vee$ indique la précomposition par la dualité $\mathrm{Hom}_k(-, k)$.

On peut également généraliser à un anneau quelconque :

Théorème 1.9 ([Dja12]). *Soient A un anneau muni d'une anti-involution et F un foncteur polynomial de $\mathbf{P}(A)$ vers les groupes abéliens. Il existe une suite spectrale*

2. Cette notion sera introduite et discutée plus tard. On peut penser par exemple à un sous-quotient d'une puissance tensorielle.

3. On a aussi un résultat analogue en caractéristique 2, mais il faut alors remplacer le groupe orthogonal par son sous-groupe d'indice 2.

naturelle

$$E^2_{p,q} = H_p(U_\infty(A); \mathrm{Tor}_q^{\mathbf{P}(A)}(\mathbb{Z}[S^2_A]^\vee, F)) \Rightarrow H_{p+q}(U_\infty(A); F_\infty)$$

qui s'effondre à la deuxième page.

Exemples de calculs explicites.

Théorème 1.10 (Djament–Vespa [DV10]). *Soit k un corps fini de cardinal q impair. L'algèbre de cohomologie stable des groupes orthogonaux sur k à coefficients dans les puissances symétriques est polynomiale sur des générateurs de bidegré $(2q^s m, q^s + 1)$ (où le premier degré est le degré homologique et le second le degré interne) indexés par les entiers naturels m et s.*

Théorème 1.11 ([Dja12]). *Soient k un corps commutatif de caractéristique nulle ; pour $n \in \mathbb{N}$, notons $\mathfrak{o}_{n,n}(k)$ la représentation adjointe de $O_{n,n}(k)$. Alors le k-espace vectoriel gradué*

$$\operatorname*{colim}_{n\in\mathbb{N}} H_*(O_{n,n}(k); \mathfrak{o}_{n,n}(k))$$

est isomorphe au produit tensoriel (sur $\mathbb{Q}$) de $H_*(O_{\infty,\infty}(k); \mathbb{Q})$ et de la partie impaire de la k-algèbre graduée $\Omega^*_k = \Lambda^*(\Omega^1_k)$ des différentielles de Kähler de k (vu comme $\mathbb{Q}$-algèbre). En particulier, cet espace vectoriel gradué est nul si k est une extension algébrique de $\mathbb{Q}$.

Il y a d'autres conséquences, plus qualitatives (changement de corps ou d'anneau etc.).

1.3. Stabilité homologique

Le problème de la stabilité homologique pour les groupes discrets fut d'abord étudié pour les seuls coefficients constants (l'un des cas les plus considérés, celui des groupes linéaires, remonte à Quillen, à qui l'on doit, semble-t-il, la première démonstration de ladite stabilité pour le cas des corps commutatifs autres que $\mathbb{F}_2$, qu'il ne publia pas). Dwyer montra en 1980 (cf. [Dwy80]), dans la situation qui l'occupait (les groupes linéaires sur des anneaux principaux), que l'on pouvait, tout en adoptant la même stratégie générale que celle toujours employée dans les raisonnements de stabilité homologique (établir la haute connectivité de complexes munis d'une action appropriée des groupes considérés, faisant apparaître les groupes précédents de la suite comme stabilisateurs), raffiner les arguments pour obtenir la stabilité à coefficients tordus favorables. Fait remarquable, les coefficients favorables dont il s'agit sont essentiellement ceux provenant de foncteurs polynomiaux en un sens approprié, coefficients qui sont aussi ceux où l'on peut obtenir les résultats les plus significatifs en matière de calculs d'homologie stable. Depuis ce travail de Dwyer, la plupart des résultats de stabilité homologique (dans le contexte qui est le nôtre) ont été traités (ou pourraient l'être sans grande difficulté) aussi bien à coefficients constants que tordus.

Nakaoka ([Nak60]) a calculé entièrement, il y a plus de cinquante ans, l'homologie de tous les **groupes symétriques**, démontrant en particulier leur stabilité homologique (à coefficients constants, mais le cas des coefficients tordus s'en déduit sans peine, comme l'a observé Betley dans [Bet02]). Dans ce cas advient aussi le phénomène exceptionnel que le morphisme canonique $H_*(\Sigma_n) \to H_*(\Sigma_{n+1})$ est en tout degré un monomorphisme scindé. Néanmoins, on peut établir la stabilité homologique pour les groupes symétriques de façon nettement plus rapide que par un calcul complet, par des méthodes analogues à (et plus simples que) celles utilisées dans les autres cas.

Les **groupes linéaires** ont reçu une attention particulière, en raison des liens avec la K-théorie algébrique et de la difficulté à étudier leur stabilité homologique sur un anneau quelconque. Celle-ci se trouve en défaut sur certains anneaux (on peut le déduire facilement de résultats de van der Kallen), mais vaut pour la plupart des anneaux raisonnables : le cas le plus général connu est celui des anneaux de rang stable de Bass fini (par exemple, les algèbres commutatives de type fini sur un corps). L'article [vdK80] de van der Kallen qui établit ce résultat demeure un tour de force technique.

La stabilité homologique pour les **groupes orthogonaux** (ou unitaires) *hyperboliques* (i.e., du type $O_{n,n}$: la stabilité pour les groupes de type euclidien O_n peut s'avérer en défaut même sur des corps commutatifs) et symplectiques vaut également sous une hypothèse analogue (à savoir : que l'anneau de base a un rang stable unitaire fini), comme l'ont montré Mirzaii et van der Kallen (cf. [MvdK02]). Pour des groupes orthogonaux non hyperboliques mais euclidiens, par exemple (cas dont nous ne parlerons pas du tout ici), la situation est plus compliquée, même sur un corps commutatif : on a besoin d'hypothèses arithmétiques (cf. [Vog82] et [Col11]).

L'homologie des **groupes d'automorphismes des groupes libres**, en lien avec des considérations de topologie différentielle, a été étudiée de façon intensive depuis les années 1980 ; Hatcher a obtenu en 1995, dans [Hat95], la stabilité homologique pour ces groupes. La borne a été améliorée et de nombreux autres résultats connexes sont apparus depuis.

Remarque 1.12. Malgré les grandes similitudes des méthodes utilisées dans les différentes situations évoquées pour établir la stabilité homologique, il semble fort ardu de dégager un cadre formel convaincant les englobant. En tout cas, il paraît raisonnable de partir d'une catégorie monoïdale *symétrique* pour obtenir des résultats généraux : les sous-groupes de congruences des groupes linéaires (qu'on peut voir comme groupes d'automorphismes de sommes itérées dans une catégorie monoïdale non symétrique) ont parfois, même sur des anneaux finis, une homologie qui ne se stabilise pas (c'est vrai par exemple pour les groupes $\mathrm{Ker}(GL_n(\mathbb{Z}/p^2) \to GL_n(\mathbb{Z}/p))$, qui sont abéliens), même s'il existe des classes intéressantes de groupes de congruences pour lesquelles vaut la stabilité. Charney a étudié cette question dans [Cha84].

1.4. Quelques résultats sur l'homologie stable à coefficients constants

Comme on l'a déjà mentionné, le calcul de l'homologie des groupes symétriques est un travail ancien de Nakaoka ([Nak60]), qui utilisait des méthodes topologiques ; dans [AM04] c'est une approche algébrique qui est suivie pour ce même calcul. Un résultat récent et difficile est le théorème remarquable suivant de Galatius (voir [Gal]) : l'inclusion $\Sigma_n \to \mathrm{Aut}\,(\mathbb{Z}^{*n})$ des groupes symétriques dans les groupes d'automorphismes des groupes libres induit stablement un isomorphisme en homologie.

L'homologie des groupes linéaires est un problème difficile et fécond (même en se restreignant à des corps commutatifs). Les premiers résultats significatifs sont dus à Quillen (cf. [Qui72]), qui a calculé entièrement, lorsque k est un corps fini de caractéristique p, la (co)homologie des groupes $GL_n(k)$ en caractéristique première à p (ce pour tout n) ; il a également démontré que $\tilde{H}_*(GL_n(\bar{\mathbb{F}}_p);\mathbb{F}_p)=0$ (encore pour tout n ; $\tilde{H}$ désigne l'homologie réduite) et que $\tilde{H}_*(GL_\infty(k);\mathbb{F}_p)=0$. En revanche, le calcul de l'homologie (instable) en caractéristique p des $GL_n(k)$ (k étant toujours fini de caractéristique p) reste très largement ouvert.

Disons quelques mots sur la méthode de Quillen pour obtenir cette nullité stable : le point remarquable est que tout repose sur la nullité de $\tilde{H}_i(GL_n(K);\mathbb{F}_p)$ pour $i\leq d$ lorsque K est un corps à p^r éléments avec $r>\frac{d}{p-1}$ (n étant là encore arbitraire !). Cette annulation repose elle-même sur l'annulation homologique analogue pour le sous-groupe des matrices triangulaires supérieures, qui s'établit par récurrence en utilisant la théorie de Galois des corps finis pour comprendre l'action du groupe multiplicatif $K^\times$ sur l'homologie du groupe additif sous-jacent à un K-espace vectoriel. Cela acquis, on utilise, pour $k\subset K$ extension finie de degré l, les inclusions $GL_n(k)\to GL_n(K)\to GL_{ln}(k)$ pour déduire l'annulation de $\tilde{H}_i(GL_\infty(k);\mathbb{F}_p)$ de celle de $\tilde{H}_i(GL_\infty(K);\mathbb{F}_p)$. Nous mentionnons cet argument non seulement pour son élégance et sa simplicité, mais aussi parce que c'est sur une variante à coefficients que reposent les premiers résultats d'annulation homologique stable pour les groupes linéaires à coefficients tordus par un foncteur polynomial sans terme constant, dus à Betley (cf. [Bet89]). L'un des intérêts des arguments d'homologie des foncteurs de Scorichenko est de démontrer ce résultat de façon totalement générale (et rapide) sans nécessiter aucun argument arithmétique.

Un résultat remarquable de Suslin (cf. [Sus83]) affirme que si $k\subset K$ est une extension de corps (commutatifs) algébriquement clos et $n>0$ un entier, l'inclusion induit un isomorphisme $H_*(GL_\infty(k);\mathbb{Z}/n)\to H_*(GL_\infty(K);\mathbb{Z}/n)$. En particulier, l'annulation stable de Quillen en caractéristique p s'étend au groupe linéaire infini sur tout corps algébriquement clos de caractéristique p.

Des résultats sont également disponibles pour l'homologie des groupes linéaires sur des corps de caractéristique nulle, et même sur certains anneaux ; on évoquera rapidement certains d'entre eux dans le paragraphe suivant.

Les résultats de Quillen sur l'homologie des groupes linéaires sur les corps finis ont été généralisés aux autres suites infinies de groupes classiques (orthogonaux, unitaires, symplectiques) sur les corps finis par Fiedorowicz et Priddy notamment (cf. [FP78]). En particulier, la trivialité stable en caractéristique naturelle (en remplaçant le groupe orthogonal par son sous-groupe d'indice 2 lorsqu'on est en caractéristique 2) vaut encore dans ce cadre. La propriété de rigidité de Suslin susmentionnée pour les extensions de corps algébriquement clos vaut encore pour les groupes orthogonaux (hyperboliques) et symplectiques, comme l'a montré Karoubi (cf. [Kar83]), à qui l'on doit plusieurs autres résultats substantiels sur l'homologie stable des groupes orthogonaux et symplectiques (cf. [Kar80]).

1.5. Relations avec d'autres théories et problèmes (co)homologiques

K-théorie algébrique. L'homologie du groupe linéaire infini sur un anneau A (à coefficients constants) est étroitement liée à la K-théorie algébrique de A, qu'on peut définir comme la suite des groupes d'homotopie de l'espace $BGL_\infty(A)^+$, la construction plus de Quillen étant relative au sous-groupe distingué parfait de $GL_\infty(A)$ engendré par les matrices élémentaires (une bonne référence pour cette construction est [Lod76] ; on peut aussi utiliser la construction catégorique Q, due également à Quillen – cf. [Qui73] –, ou la construction de Volodin, etc.). L'une des motivations de l'introduction de cet espace est de transcrire homotopiquement la structure d'algèbre de Hopf sur $H^*(GL_\infty(A))$: en effet, les morphismes $GL_n \times GL_m \to GL_{n+m}$ qui induisent cette structure ne font du classifiant de $GL_\infty(A)$ un H-espace qu'une fois appliquée la construction plus.

Plus directement reliée aux considérations de coefficients tordus dont on traite dans ce cours, la *K-théorie stable* de l'anneau A à coefficients dans un $GL(A)$-module (ici $GL := GL_\infty$) M : c'est par définition l'homologie de la fibre homotopique de l'application canonique $BGL(A) \to BGL(A)^+$ à coefficients tordus par M ; on la note $K_*^s(A;M)$. L'observation essentielle est que, dans la suite spectrale de Serre

$$E^2_{p,q} = H_p(GL(A);K_q^s(A;M)) = H_p(BGL(A)^+;K_q^s(A;M)) \Rightarrow H_{p+q}(GL(A);M)$$

l'action du groupe linéaire sur la K-théorie stable est *triviale*.

Nous n'en dirons pas davantage à ce sujet, car nous ne privilégierons pas la présentation des résultats de Scorichenko en termes de K-théorie stable, la version reliant directement l'homologie stable des groupes linéaires à coefficients tordus à l'homologie stable à coefficients constants et l'homologie de Hochschild de la catégorie $\mathbf{P}(A)$ nous semblant plus adaptée ici (de même, nous pourrions parler de K-théorie hermitienne stable pour évoquer les résultats sur l'homologie stable des groupes unitaires).

Conjecture de Friedlander–Milnor. Elle affirme que si G est un groupe de Lie et G^δ désigne le même groupe muni de la topologie discrète, l'application canonique $H_*(BG^\delta;\mathbb{Z}/n) \to H_*(BG;\mathbb{Z}/n)$ est un isomorphisme pour entier $n > 0$. Le membre de droite est raisonnablement calculable, alors que celui de gauche est extrêmement difficile à estimer. Dans [Mil83], Milnor discute cette conjecture et la démontre dans le cas élémentaire des groupes résolubles.

Suslin a démontré que cette conjecture est valide stablement pour les groupes linéaires sur $\mathbb{R}$ ou $\mathbb{C}$ (cf. [Sus87]). Des progrès majeurs vers une démonstration complète de la conjecture ont récemment été réalisés par Morel.

Cohomologie des groupes algébriques, cohomologie continue. La conjecture de Friedlander–Milnor tombe trivialement en défaut lorsqu'on remplace les coefficients finis par $\mathbb{Q}$ [4]. Néanmoins, de nombreux résultats existent sur l'homologie rationnelle des groupes de Lie rendus discrets et de sous-groupes remarquables ; c'est en un sens, selon l'appendice de [Mil83], une situation plus facile à traiter que celle des coefficients

---

4. Cf. par exemple le groupe de Lie additif $\mathbb{R}$, dont l'homologie rationnelle comme groupe discret est énorme – algèbre à puissances extérieures de $\mathbb{R}$ vu comme $\mathbb{Q}$-espace vectoriel – tandis que le classifiant du groupe de Lie est homologiquement trivial.

finis. L'outil essentiel est de relier la cohomologie réelle d'un groupe de Lie rendu discret à la cohomologie *continue* du même groupe de Lie. On doit notamment à Borel des résultats très substantiels en la matière, qui permettent de calculer par exemple $H_*(GL_\infty(k);\mathbb{Q})$ lorsque k est le corps des réels, mais aussi un corps de nombres.

Une sorte d'analogue en caractéristique finie est le lien établi par Cline, Parshall, Scott et van der Kallen entre la cohomologie d'un groupe algébrique (semi-simple et déployé) sur $\mathbb{F}_p$, à coefficients dans une représentation rationnelle, après l'avoir suffisamment tordue par le morphisme de Frobenius, et la cohomologie du groupe discret des points sur $\mathbb{F}_{p^d}$, avec d assez grand, de cette représentation (cf. [CPSK77]). Du reste, les liens profonds entre (co)homologie des foncteurs polynomiaux et (co)homologie des groupes (discrets) d'automorphismes correspondants dont on traite dans ce cours possèdent un pendant en termes de groupes algébriques, les foncteurs polynomiaux devant être remplacés par les foncteurs *strictement* polynomiaux (possédant eux-mêmes des liens étroits avec les foncteurs polynomiaux ordinaires : ainsi, dans [Bet99], Betley utilise les foncteurs strictement polynomiaux pour démontrer l'isomorphisme entre homologie des foncteurs et homologie stable des groupes linéaires (discrets) sur un corps fini à coefficients polynomiaux). Nous tairons dans la suite cet aspect (voir [Tou10]), que d'autres exposés de cette semaine sur les foncteurs aborderont.

Groupes de découpage. L'homologie des groupes de Lie rendus discrets (mais pas dans le domaine stable, cette fois-ci, et avec des coefficients tordus appropriés) apparaît naturellement dans l'étude des groupes de découpage et du troisième problème de Hilbert. On renvoie le lecteur à l'ouvrage de Dupont [Dup01] à ce sujet.

2. Premier lien entre homologie stable des groupes discrets et homologie des foncteurs

Abstract. *The second lecture presents a first isomorphism result between stable group homology with twisted coefficients and functor homology. It works in a rather general setting, and it makes no assumption on the functor. Its proof is therefore elementary. Unfortunately, but for the case of the symmetric groups (which is quickly studied), the functor homology which appears is very hard to compute. The following lectures will compare this functor homology to a much more manageable functor homology, thus leading to the introduction's statements.*

Le but de cette section est de donner un premier résultat d'isomorphisme entre homologie stable de groupe à coefficients tordus et homologie des foncteurs (modulo la connaissance de l'homologie stable de groupe à coefficients constants) dans un cadre assez général, sans aucune hypothèse sur le foncteur par lequel on tord les coefficients. De ce fait, la démonstration en est formelle et élémentaire ; en revanche, hors du cas particulièrement favorable des groupes symétriques, on n'obtient guère de résultat directement exploitable, car l'homologie de foncteurs qui apparaît est assez inaccessible. C'est ensuite la comparaison de cette homologie de foncteurs à d'autres groupes d'homologie de foncteurs nettement plus exploitables qui permettra d'obtenir les théorèmes annoncés dans l'introduction (mais cela nécessitera davantage d'hypothèses et beaucoup plus de travail : ce sera l'objet des trois dernières sections).

Tous les résultats de cette partie sont tirés de [DV10] (plus précisément, de sa première section, sauf les considérations sur les groupes symétriques traitées dans un appendice de cet article).

Dans toute cette section, $\Bbbk$ est un anneau commutatif de base fixé (ce sera en général soit $\mathbb{Z}$, soit un corps) et $(\mathcal{C}, \oplus, 0)$ est une petite catégorie[5] monoïdale symétrique dont l'unité 0 est objet initial, X est un objet de $\mathcal{C}$ et l'on s'intéresse aux groupes $G_n := \mathrm{Aut}_{\mathcal{C}}(X^{\oplus n})$. On cherche à relier leur homologie stable, i.e., l'homologie de leur colimite (sur n), à l'homologie de la catégorie $\mathcal{C}$. Comme $\mathcal{C}$ a un objet initial, l'homologie de $\mathcal{C}$ à coefficients constants est triviale (le foncteur constant $\Bbbk$ de $\mathcal{C}$-**Mod** est projectif – ou, d'un point de vue topologique, le classifiant de cette catégorie est contractile) : c'est la situation à coefficients tordus qui est intéressante.

2.1. Rappels élémentaires d'algèbre homologique dans la catégorie $\mathcal{C}$-Mod des foncteurs de $\mathcal{C}$ vers $\Bbbk$-Mod

La catégorie $\mathcal{C}$-**Mod** est une catégorie abélienne agréable : elle possède des limites et colimites quelconques, qui se calculent au but ; les colimites filtrantes sont exactes. De plus, elle possède un *ensemble* de générateurs projectifs distingués : $P_A^{\mathcal{C}} := \Bbbk[\mathrm{Hom}_{\mathcal{C}}(A, -)]$; en effet, le lemme de Yoneda fournit un isomorphisme canonique $\mathrm{Hom}\,(P_A^{\mathcal{C}}, F) \simeq F(A)$. Cela permet de faire de l'algèbre homologique dans $\mathcal{C}$-**Mod** (qui possède aussi assez d'objets injectifs) comme dans des catégories de modules sur un anneau. En particulier, on dispose de la notion de *groupes de torsion* sur $\mathcal{C}$, qui s'obtiennent en dérivant à gauche le bifoncteur $- \underset{\mathcal{C}}{\otimes} - : (\mathbf{Mod}\text{-}\mathcal{C}) \times (\mathcal{C}\text{-}\mathbf{Mod}) \to \Bbbk\text{-}\mathbf{Mod}$ qu'on peut caractériser par sa commutation aux colimites *relativement à chaque variable* et des isomorphismes canoniques $P_A^{\mathcal{C}^{\mathrm{op}}} \underset{\mathcal{C}}{\otimes} F \simeq F(A)$, $G \underset{\mathcal{C}}{\otimes} P_A^{\mathcal{C}} \simeq G(A)$, ou par l'adjonction

$$\mathrm{Hom}_{\Bbbk}(G \underset{\mathcal{C}}{\otimes} F, M) \simeq \mathrm{Hom}_{\mathcal{C}\text{-}\mathbf{Mod}}(F, \mathcal{H}om(G, M))$$

où $\mathcal{H}om(G, M) := \mathrm{Hom}_{\Bbbk}(-, M) \circ G$.

L'*homologie* de $\mathcal{C}$ à coefficients dans $F \in \mathrm{Ob}\,\mathcal{C}$-**Mod** est par définition $H_*(\mathcal{C}; F) := \mathrm{Tor}_*^{\mathcal{C}}(\Bbbk, F)$ (où $\Bbbk$ désigne le foncteur constant en $\Bbbk$ de **Mod**-$\mathcal{C}$).

On dispose également d'une fonctorialité en la catégorie $\mathcal{C}$: si $\Phi : \mathcal{C} \to \mathcal{D}$ est un foncteur entre petites catégories, Φ induit un foncteur exact de précomposition $\Phi^* : \mathcal{D}\text{-}\mathbf{Mod} \to \mathcal{C}\text{-}\mathbf{Mod}$ et on dispose, pour $F \in \mathrm{Ob}\,\mathcal{D}$-**Mod**, d'un morphisme naturel $H_*(\mathcal{C}; \Phi^* F) \to H_*(\mathcal{D}; F)$ vérifiant de nombreuses propriétés de cohérence.

5. Ou plus généralement, essentiellement petite.

2.2. Les morphismes naturels $H_*(G_\infty; F_\infty) \to H_*(\mathcal{C}; F)$ et $H_*(G_\infty; F_\infty) \to H_*(G_\infty \times \mathcal{C}; F)$

Pour tout $n \in \mathbb{N}$, on a un foncteur évident d'image $X^{\oplus n}$ de la catégorie à un objet dont les flèches sont données par le groupe G_n (qu'on notera encore G_n) vers la catégorie $\mathcal{C}$: elle induit un morphisme naturel de $\Bbbk$-modules gradués $H_*(G_n; F(X^{\oplus n})) \to H_*(\mathcal{C}; F)$. Le diagramme

$$\begin{array}{ccc} H_*(G_n; F(X^{\oplus n})) & \longrightarrow & H_*(G_{n+1}; F(X^{\oplus n+1})) \\ \downarrow & \swarrow & \\ H_*(\mathcal{C}; F) & & \end{array}$$

commute (parce qu'il y a une transformation naturelle de $G_n \to \mathcal{C}$ vers $G_n \to G_{n+1} \to \mathcal{C}$, donnée par le morphisme canonique $A \to A \oplus X$), de sorte qu'on obtient un morphisme naturel $H_*(G_\infty; F_\infty) \to H_*(\mathcal{C}; F)$.

On aimerait qu'il soit un isomorphisme (sous de bonnes hypothèses), mais une première obstruction fâcheuse vient du cas $F = \Bbbk$ (foncteur constant) : l'homologie $H_*(\mathcal{C}; \Bbbk)$ étant triviale, on ne peut s'attendre à ce que la flèche précédente le soit pour tout F (raisonnable) que si $H_*(G_\infty; \Bbbk)$ est également triviale ! Cela arrive dans un certain nombre de cas importants qui nous intéressent (par exemple : les groupes linéaires sur un corps fini k, avec $\Bbbk = k$), mais pas tous, on va donc faire en sorte de tenir compte « artificiellement » de l'homologie stable à coefficients constants des G_n, que les méthodes d'homologie des foncteurs développées dans ce cours ne permettent pas d'atteindre. Pour cela, on considère le foncteur de projection $\Pi : G_\infty \times \mathcal{C} \to \mathcal{C}$ et l'homologie $H_*(G_\infty \times \mathcal{C}; \Pi^* F)$, notée par abus, $H_*(G_\infty \times \mathcal{C}; F)$: elle donne le bon résultat pour F constant, et nous verrons qu'elle se calcule simplement à partir de $H_*(G_\infty; \Bbbk)$ et de $H_*(\mathcal{C}; F)$ (cf. paragraphe suivant).

On utilise donc les foncteurs $G_n \to G_\infty \times \mathcal{C}$ dont la composante $G_n \to G_\infty$ est le morphisme canonique et $G_n \to \mathcal{C}$ le foncteur utilisé précédemment : cela procure des morphismes naturels $H_*(G_n; F(X^{\oplus n})) \to H_*(G_\infty \times \mathcal{C}; F)$ puis $H_*(G_\infty; F_\infty) \to H_*(G_\infty \times \mathcal{C}; F)$.

2.3. Lien entre $H_*(\mathcal{C}; F)$, $H_*(G_\infty; \Bbbk)$ et $H_*(G_\infty \times \mathcal{C}; F)$

On renvoie à [DV10], ğ 2.3, où ces questions sont traitées en détail. On dispose de deux suites spectrales qui aboutissent à $H_*(G_\infty \times \mathcal{C}; F)$; elles dégénèrent souvent à la deuxième page. Les deux propriétés suivantes (qui relèvent de l'algèbre homologique classique) nous suffiront amplement :

Proposition 2.1. *Si $\Bbbk$ est un corps, alors on dispose d'un isomorphisme naturel*

$$H_*(G_\infty \times \mathcal{C}; F) \simeq H_*(G_\infty; \Bbbk) \underset{\Bbbk}{\otimes} H_*(\mathcal{C}; F)$$

de $\Bbbk$-modules gradués.

Proposition 2.2. *Si $\Bbbk = \mathbb{Z}$, alors on a un isomorphisme naturel gradué*

$$H_*(G_\infty \times \mathcal{C}; F) \simeq \mathrm{Tor}_*^{\mathcal{C}}(H_*(G_\infty; \mathbb{Z}), F)$$

où $H_*(G_\infty; \mathbb{Z})$ est vu comme un foncteur (contravariant) constant sur $\mathcal{C}$.

Si M est un groupe abélien (vu comme foncteur constant), on dispose de suites exactes naturelles scindées (en général non naturellement)

$$0 \to M \underset{\mathbb{Z}}{\otimes} H_p(\mathcal{C}; F) \to \mathrm{Tor}_p^{\mathcal{C}}(M, F) \to \mathrm{Tor}_1^{\mathbb{Z}}(M, H_{p-1}(\mathcal{C}; F)) \to 0$$

de groupes abéliens.

2.4. Les hypothèses pertinentes sur $(\mathcal{C}, \oplus, 0)$ et X

1. (Transitivité stable) Pour tout morphisme $f : A \to B$ de $\mathcal{C}$, il existe un automorphisme α de $A \oplus B$ faisant commuter le diagramme suivant.

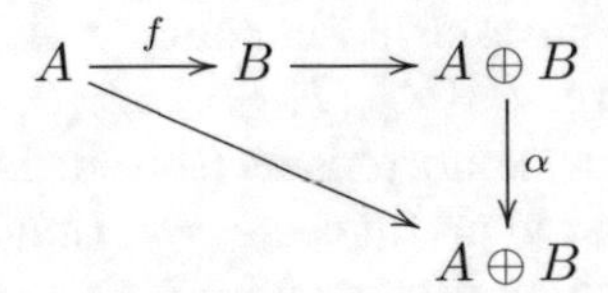

(On pourrait exiger un peu moins encore, mais cette propriété sera vérifiée dans tous les cas qui nous intéressent.)

Dans certains cas, l'axiome plus fort suivant sera vérifié :

(Version forte : transivité instable) pour tous objets A et B de $\mathcal{C}$, l'action du groupe $\mathrm{Aut}_{\mathcal{C}}(B)$ sur l'ensemble $\mathrm{Hom}_{\mathcal{C}}(A, B)$ est transitive.

2. (Stabilisateurs) Étant donné des objets A et B de $\mathcal{C}$, le morphisme de groupes canonique $\mathrm{Aut}_{\mathcal{C}}(B) \to \mathrm{Aut}_{\mathcal{C}}(A \oplus B)$ est une injection dont l'image est le sous-groupe des automorphismes φ faisant commuter le diagramme suivant.

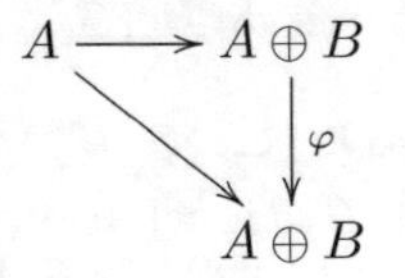

3. (Engendrement faible par X) pour tout objet A de $\mathcal{C}$, il existe un objet B de $\mathcal{C}$ et un entier naturel n tels que $A \oplus B \simeq X^{\oplus n}$

Dans certains cas, l'axiome suivant sera même vérifié :

(Engendrement fort par X) tout objet de $\mathcal{C}$ est isomorphe à un $X^{\oplus n}$.

2.5. Les exemples fondamentaux

1. (Groupes symétriques) $\mathcal{C}$ est la catégorie Θ des ensembles finis *avec injections*, la structure monoïdale est la réunion disjointe, on prend pour X un ensemble à un élément. Toutes les hypothèses précédentes sont satisfaites dans ce cas, comme on le vérifie aussitôt.
2. (Groupes linéaires) Soient A un anneau et $\mathbf{P}(A)$ la catégorie des A-modules à gauche projectifs de type fini, qu'on munit de la somme directe. Cette catégorie ne vérifie évidemment pas les propriétés requises. La sous-catégorie

des monomorphismes scindés $\mathbf{M}(A)$ vérifie l'hypothèse de transitivité stable (et bien sûr celle d'engendrement), mais pas celle relative aux stabilisateurs. Pour y remédier, on considère une troisième catégorie (qui nous permet de simplifier les arguments de Scorichenko), notée $\mathbf{S}(A)$, qui a toujours les mêmes objets, mais dont les morphismes $M \to N$ sont les couples (v, u) formés d'applications A-linéaires $v : N \to M$ et $u : M \to N$ telles que $v \circ u = Id_M$. Cette fois-ci toutes les propriétés sont satisfaites avec $X = A$ (la transitivité n'est toutefois en général pas vraie sous forme instable, sauf dans des cas où l'anneau A est particulièrement gentil, par exemple un corps, tout comme l'engendrement fort).

3. (Groupes unitaires) Soient A un anneau muni d'une anti-involution, $\epsilon \in \{-1, 1\}$ et $\mathbf{H}_\epsilon(A)$ (on omet par abus toute mention à l'anti-involution) la catégorie des A-modules à gauche projectifs de type fini ϵ-hermitiens non dégénérés. Alors toutes les propriétés précédentes sont vérifiées avec pour X l'espace hermitien hyperbolique standard A^2 – là encore, la transitivité instable n'est généralement pas vraie (elle l'est cependant si A est un corps par le théorème de Witt), et l'engendrement (qui vient de ce que, si (M, q) est un espace hermitien non dégénéré, $(M, q) \overset{\perp}{\oplus} (M, -q)$ est hyperbolique) n'est vrai que sous forme faible. Il est indispensable de considérer des espaces non dégénérés !

   La situation précédente est en fait un cas particulier de celle-ci, appliquée à l'anneau $A^{\mathrm{op}} \times A$ muni de l'anti-involution $(a, b) \mapsto (b, a)$ (avec $\epsilon = 1$).

2.6. Résultat principal

Proposition 2.3 ([DV10]). *Sous les trois hypothèses précédentes, le morphisme naturel de $\Bbbk$-modules gradués*

$$H_*(G_\infty; F_\infty) \to H_*(G_\infty \times \mathcal{C}; F)$$

est un isomorphisme pour tout $F \in \mathrm{Ob}\,\mathcal{C}$-$\mathbf{Mod}$. En particulier, si $\tilde{H}_*(G_\infty; \Bbbk) = 0$, alors le morphisme $H_*(G_\infty; F_\infty) \to H_*(\mathcal{C}; F)$ est un isomorphisme.

Démonstration. Quitte à remplacer $\mathcal{C}$ par la sous-catégorie pleine des $X^{\oplus n}$, on se ramène aisément au cas où l'hypothèse d'engendrement *fort* est satisfaite.

Il suffit de démontrer cette assertion lorsque F est un foncteur projectif standard $P^{\mathcal{C}}_A$ (résoudre un foncteur arbitraire par des sommes directes de projectifs standard et comparer les suites spectrales). On se donne un entier m tel que $A \simeq X^{\oplus m}$ (hypothèse d'engendrement).

Comme $\tilde{H}_*(\mathcal{C}; P^{\mathcal{C}}_A) = 0$, $H_*(G_\infty \times \mathcal{C}; P^{\mathcal{C}}_A) \simeq H_*(G_\infty; \Bbbk)$. Par ailleurs, le lemme de Shapiro donne

$$H_*(G_n; P^{\mathcal{C}}_A(X^{\oplus n})) \simeq \bigoplus_{\text{orbites}} H_*(\mathrm{Stab}; \Bbbk)$$

où la somme est prise sur les orbites de l'action de G_n sur l'ensemble $\mathrm{Hom}_{\mathcal{C}}(A, X^{\oplus n})$ et Stab désigne le stabilisateur d'un représentant de l'orbite. De plus, avec cette description, le morphisme $H_*(G_n; P_A^{\mathcal{C}}(X^{\oplus n})) \to H_*(G_\infty \times \mathcal{C}; P_A^{\mathcal{C}})$ qui nous intéresse a pour composantes les morphismes $H_*(\mathrm{Stab}, \Bbbk) \to H_*(G_\infty; \Bbbk)$ induits par les morphismes canoniques $\mathrm{Stab} \subset G_n \to G_\infty$.

Montrons la surjectivité de notre morphisme. Pour cela, on établit que, pour tout $i \in \mathbb{N}$, l'image de $H_*(G_i) \to H_*(G_\infty)$ (tous les coefficients étant pris dans $\Bbbk$) est incluse dans son image. Pour cela, il suffit de considérer, pour $n = m + i$, le morphisme $v : A \simeq X^{\oplus m} \to X^{\oplus m} \oplus X^{\oplus i} = X^{\oplus n}$. Son stabilisateur sous l'action de G_n contient l'image de G_i (ici on regarde une flèche $G_i \to G_n$ qui n'est pas celle de départ, mais lui est conjuguée par un automorphisme de permutation des facteurs, donc induit le même morphisme en homologie), ce qui établit notre assertion.

Montrons maintenant l'injectivité. L'hypothèse de transitivité stable montre que le morphisme $G_n \to G_{n+m}$ induit en homologie à coefficients tordus par $P_A^{\mathcal{C}}$ une application dont l'image est incluse (via la décomposition précédente) dans l'homologie du stabilisateur du morphisme canonique $A \simeq X^{\oplus m} \to X^{\oplus m+n}$. Mais ce stabilisateur n'est autre que l'image de G_n dans G_{m+n}, par hypothèse. On en déduit l'injectivité stable de nos applications, ce qui termine la démonstration. □

La suite de ce cours consiste à expliquer comment calculer, dans les cas favorables, l'homologie $H_*(\mathcal{C}; F)$ (on suppose $H_*(G_\infty; \Bbbk)$ connue). De fait, les catégories vérifiant les axiomes utilisés sont fort peu maniables pour le calcul homologique direct. On cherche donc à se ramener à d'autres catégories, par exemple $\mathbf{P}(A)$ plutôt que $\mathbf{S}(A)$ pour étudier l'homologie des groupes linéaires sur A. Cela demandera un peu de préparation, exposée dans la prochaine section. Il y a néanmoins un cas beaucoup plus facile à aborder, qu'on évoque maintenant rapidement.

2.7. Le cas des groupes symétriques

La catégorie «ensembliste» (i.e., dont les groupes d'automorphismes sont les groupes symétriques) dans laquelle le plus de calculs d'algèbre homologique fructueux ont été menés n'est pas la catégorie Θ des ensembles finis avec injections à laquelle on peut directement appliquer la Proposition 2.3, mais la catégorie Γ des ensembles finis *pointés* (les morphismes étant toutes les fonctions envoyant le point de base de la source sur le point de base du but) – voir notamment l'article [Pir00b] de Pirashvili. En fait, on peut relier assez facilement, du point de vue homologique, les deux catégories. La tâche s'avère nettement plus aisée que dans les cas que l'on étudiera ultérieurement car *tous* les Γ-modules (foncteurs de Γ vers les $\Bbbk$-modules) sont analytiques, c'est-à-dire colimite d'objets polynomiaux, il n'y a donc pas à utiliser d'outil homologique spécifique à ces objets : des techniques d'algèbre homologique directes suffisent (on peut se contenter de regarder ce qui se passe sur les projectifs standard).

Si F est un Γ-module et i un entier naturel, on définit l'effet croisé $cr_i(F)$ par

$$cr_i(F) := \mathrm{Ker}\,(F([i]) \to F([i-1])^{\oplus i})$$

(cf. [Pir00b] par exemple), où $[i] := \{0, 1, \ldots, i\}$ (avec 0 pour point de base) et les applications $F([i]) \to F([i-1])$ sont induites par les morphismes $r_j : [i] \to [i-1]$

($j = 1, \dots, i$) envoyant 0 et j sur 0, i sur i pour $0 < i < j$ et sur $i-1$ pour $j < i \leq n$. Le $\Bbbk$-module $cr_i(F)$ est muni d'une action naturelle du groupe symétrique Σ_i.

Théorème 2.4 (Betley). *Il existe un isomorphisme naturel*

$$H_*(\Sigma_\infty; F_\infty) \simeq \bigoplus_{i \in \mathbb{N}} H_*(\Sigma_\infty \times \Sigma_i; cr_i(F))$$

de $\Bbbk$-modules gradués pour tout Γ-module F (où Σ_∞ agit trivialement sur $cr_i(F)$).

Ce résultat, démontré de manière directe par Betley dans [Bet02], est déduit de la Proposition 2.3 dans l'appendice E de [DV10], à l'aide de résultats de décomposition sur les Γ-modules dus à Pirashvili (voir [Pir00a]) élémentaires mais très utiles et d'arguments d'adjonction. Ces résultats de décomposition se traduisent d'ailleurs par l'équivalence entre la catégorie des Γ-modules et celle des Ω-modules, où Ω est la catégorie des ensembles finis (non pointés) *avec surjections*, ce qui illustre l'un des principaux mots d'ordre de ce cours : ne jamais hésiter à transiter par de multiples catégories de foncteurs !

3. Foncteurs polynomiaux ; les résultats d'annulation de Scorichenko

Abstract. *In this lecture, polynomial functors on an additive category are defined. Scorichenko's criterion for cancellation of torsion groups, one of whose argument is polynomial, is then established. The key point is to consider cross-effects as natural transformations.*

Dans cette séance, $\mathcal{A}$ désigne une petite catégorie additive[6] (c'est-à-dire qu'elle possède des sommes et produits finis et qu'ils coïncident, on dispose donc d'un objet nul et d'une notion de somme directe – mais on ne suppose nullement qu'existent noyaux ou coyonaux). On suit fidèlement le travail (non publié) [Sco00], aux notations près.

3.1. Effets croisés et foncteurs polynomiaux

Si $f : A \to B$ est une fonction entre groupes abéliens, on dit que f est *polynomiale* de degré au plus $d \in \mathbb{N}$ si, pour toute famille $(x_0, \dots, x_d)$ d'éléments de A, on a

$$\sum_{I \subset \{0,\dots,d\}} (-1)^{|I|} f(x_I) = 0$$

où $|I|$ désigne le cardinal de I et l'on note $x_I = \sum_{i \in I} x_i$. (Si $A = B = \mathbb{Z}$, par exemple, cela équivaut à dire que f est la fonction associée à un polynôme de degré au plus d – toutefois, cette remarque ne s'étend pas à n'importe quel groupe abélien source A.) Il est exactement équivalent d'exiger que

$$\sum_{I \in \mathcal{P}(E)} (-1)^{|I|} f(x_I) = 0$$

6. On peut considérer des foncteurs polynomiaux dans un cadre plus général, mais nous n'en aurons pas usage ; les arguments ici présentés sont spécifiques à cette situation, même si on peut les modifier convenablement pour traiter d'autres situations.

pour toute famille (x_i) d'éléments de A, où E est un ensemble *pointé* à $d + 2$ éléments (point de base inclus) et $\mathcal{P}(E)$ désigne l'ensemble des sous-ensembles pointés de E.

On donne maintenant un analogue fonctoriel de cette définition, en partant de la variante pointée (qui n'est pas la plus habituelle mais sera légèrement plus commode d'un point de vue technique pour nos considérations ultérieures). Les notions d'effets croisés et de foncteurs polynomiaux qu'on va introduire sont anciennes : elles remontent à Eilenberg–Mac Lane [EML54], dans les années 1950. Néanmoins, l'étude systématique des foncteurs polynomiaux, spécialement de leurs propriétés homologiques, a connu un essor plus tardif (à partir des années 1980 sans doute) et n'a probablement pas encore livré tous ses secrets.

Supposons que E est un ensemble fini. On dispose d'un foncteur $t_E : \mathcal{A} \to \mathcal{A} \quad A \mapsto A^{\oplus E}$; en fait cette construction est également fonctorielle en E. En particulier, si I est un sous-ensemble de E, on dispose de transformations naturelles $u_I : Id \to t_E$ et $p_I : t_E \to Id$ dont les composantes (évaluées sur un objet A) sont l'identité $A \to A$ pour les indices appartenant à I et 0 ailleurs. Si F est un objet de $\mathcal{A}$-**Mod**, on note $T_E(F) = t_E^* F (= F \circ t_E)$; si l'on suppose que E est un ensemble fini *pointé*, on dispose de transformations naturelles

$$cr_E^{\mathcal{A},\mathrm{dir}}(F) = \sum_{I \in \mathcal{P}(E)} (-1)^{|I|} F(u_I) : F \to T_E(F)$$

(effet croisé direct (pointé) ; les exposants seront omis s'il n'y a pas de confusion possible) et

$$cr_E^{\mathcal{A},\mathrm{inv}}(F) = \sum_{I \in \mathcal{P}(E)} (-1)^{|I|} F(p_I) : T_E(F) \to F$$

(effet croisé inverse (pointé) ; les exposants seront encore souvent omis).

Définition 3.1. On dit qu'un foncteur $F : \mathcal{A} \to \Bbbk$-**Mod** est *polynomial* de degré au plus d s'il vérifie les conditions équivalentes suivantes :

1. $cr_E^{\mathcal{A},\mathrm{dir}}(F) = 0$ pour tout ensemble pointé E de cardinal au moins $d + 2$;
2. $cr_E^{\mathcal{A},\mathrm{dir}}(F) = 0$ pour un ensemble pointé E de cardinal $d + 2$;
3. $cr_E^{\mathcal{A},\mathrm{inv}}(F) = 0$ pour tout ensemble pointé E de cardinal au moins $d + 2$;
4. $cr_E^{\mathcal{A},\mathrm{inv}}(F) = 0$ pour un ensemble pointé E de cardinal $d + 2$.

On dit que F est *analytique* s'il est colimite de foncteurs polynomiaux (si c'est le cas, on peut écrire F comme la colimite filtrante de sous-foncteurs polynomiaux).

Exemple 3.2. Les foncteurs polynomiaux de degré 0 sont exactement les foncteurs constants. À tout foncteur F correspond une décomposition canonique $F \simeq F(0) \oplus \bar{F}$ (où $F(0)$ est vu comme un foncteur constant), puisque 0 est objet nul de $\mathcal{A}$; le foncteur F est polynomial de degré au plus 1 si et seulement si $\bar{F}$ est additif au sens où le morphisme canonique $\bar{F}(A) \oplus \bar{F}(B) \to \bar{F}(A \oplus B)$ est un isomorphisme pour tous objets A et B de $\mathcal{A}$.

Remarque 3.3.

1. La classe des foncteurs polynomiaux de degré au plus d jouit de nombreuses propriétés de stabilité : elle est préservée par limites et colimites quelconques, stable par sous-quotients et par extensions.
2. Un produit tensoriel de foncteurs polynomiaux est polynomial et son degré est au plus la somme des degrés des foncteurs d'origine.
3. Lorsque cela fait sens, on vérifie également qu'une composition de foncteurs polynomiaux est polynomiale, le degré étant au plus le produit des degrés initiaux.
4. Les exemples typiques de foncteurs polynomiaux de degré d depuis une catégorie de modules sont les d-èmes puissances tensorielles, symétriques, divisées, extérieures... De plus, la puissance tensorielle T^d, munie de son action canonique du groupe symétrique Σ_d, joue un rôle essentiel dans la classification des foncteurs polynomiaux de degré au plus d modulo ceux de degré strictement inférieur, au moins lorsque la source est une catégorie de modules très sympathique – par exemple, la catégorie des espaces vectoriels de dimension finie sur un corps premier.
5. De nombreux foncteurs ne sont pas analytiques ; l'exemple 3.8 ci-après fournira même un foncteur (non nul) n'ayant aucun quotient analytique non nul.
6. On peut définir une notion de foncteur polynomial dans un cadre plus général que celui ici abordé (source additive), mais les résultats d'annulation homologique de Scorichenko exposés dans la suite de cette section ne s'appliquent essentiellement qu'à cette situation.

3.2. Adjonction entre effets croisés, application aux groupes de torsion

Le fait que $\mathcal{A}$ est une catégorie additive implique que, pour tout ensemble fini E, l'endofoncteur t_E de $\mathcal{A}$ est adjoint à lui-même : $\mathrm{Hom}_{\mathcal{A}}(A, t_E(B)) \simeq \mathrm{Hom}_{\mathcal{A}}(t_E(A), B)$ naturellement en A et B (ces groupes abéliens étant tous deux naturellement isomorphes à $\mathrm{Hom}_{\mathcal{A}}(A,B)^E$). Via cette adjonction, les transformations naturelles u_I et p_I (où I est une partie de E) se correspondent, en ce sens que le diagramme

$$\begin{array}{ccc} \mathrm{Hom}_{\mathcal{A}}(A, t_E(B)) & \xrightarrow{\simeq} & \mathrm{Hom}_{\mathcal{A}}(t_E(A), B) \\ \downarrow{\scriptstyle \mathrm{Hom}_{\mathcal{A}}(A, p_I)} & & \downarrow{\scriptstyle \mathrm{Hom}_{\mathcal{A}}(u_I, B)} \\ \mathrm{Hom}_{\mathcal{A}}(A, B) & \xrightarrow{=} & \mathrm{Hom}_{\mathcal{A}}(A, B) \end{array}$$

commute (de même qu'un analogue avec les flèches verticales dans l'autre sens).

Pour des raisons formelles, ces propriétés se transportent dans les catégories de foncteurs $\mathcal{A}$-**Mod** et **Mod**-$\mathcal{A} = \mathcal{A}^{\mathrm{op}}$-**Mod** (noter que dans la catégorie additive $\mathcal{A}^{\mathrm{op}}$, les rôles de u_I et p_I sont intervertis par rapport à $\mathcal{A}$) : pour $X \in \mathrm{Ob}\,\mathbf{Mod}\text{-}\mathcal{A}$ et $F \in \mathrm{Ob}\,\mathcal{A}\text{-}\mathbf{Mod}$, on dispose d'un isomorphisme naturel

$$\mathrm{Tor}_*^{\mathcal{A}}(X, T_E(F)) \simeq \mathrm{Tor}_*^{\mathcal{A}}(T_E(X), F)$$

de $\Bbbk$-modules gradués et d'un diagramme commutatif, si E est maintenant un ensemble fini pointé,

$$\begin{array}{ccc} \mathrm{Tor}_*^{\mathcal{A}}(X, T_E(F)) & \xrightarrow{\simeq} & \mathrm{Tor}_*^{\mathcal{A}}(T_E(X), F) \\ {\scriptstyle \mathrm{Tor}_*^{\mathcal{A}}(X, cr_E^{\mathcal{A},\mathrm{inv}}(F))} \downarrow & \swarrow {\scriptstyle \mathrm{Tor}_*^{\mathcal{A}}(cr_E^{\mathcal{A}^{\mathrm{op}},\mathrm{inv}}(X), F)} & \\ \mathrm{Tor}_*^{\mathcal{A}}(X, F) & & \end{array}$$

On note par ailleurs $\kappa_E(X) = \mathrm{Ker}\, cr_E^{\mathcal{A}^{\mathrm{op}},\mathrm{inv}}(X)$ et κ_E^n, pour $n \in \mathbb{N}$, la n-ième itération de cet endofoncteur de **Mod**-$\mathcal{A}$. On désigne aussi par $\mathcal{K}_d$ la classe des $X \in \mathrm{Ob}\,\mathbf{Mod}\text{-}\mathcal{A}$ tels que $cr_E^{\mathcal{A}^{\mathrm{op}},\mathrm{inv}}(X)$ soit un épimorphisme, où E est un ensemble pointé de cardinal $d + 2$.

Proposition 3.4. *Soient $d \in \mathbb{N}$, E un ensemble fini pointé de cardinal $d + 2$, $F \in \mathrm{Ob}\,\mathcal{A}\text{-}\mathbf{Mod}$ polynomial de degré au plus d et $X \in \mathrm{Ob}\,\mathbf{Mod}\text{-}\mathcal{A}$ tel que $\kappa_E^n(X)$ appartienne à $\mathcal{K}_d$ pour tout entier n. Alors*

$$\mathrm{Tor}_*^{\mathcal{A}}(X, F) = 0.$$

Démonstration. On procède par récurrence sur le degré homologique. La suite exacte courte $0 \to \kappa_E(X) \to T_E(X) \xrightarrow{cr_E(X)} X \to 0$ ($cr_E(X)$ est par hypothèse surjectif) induit une suite exacte longue en homologie :

$$\cdots \to \mathrm{Tor}_n^{\mathcal{A}}(\kappa_E(X), F) \to \mathrm{Tor}_n^{\mathcal{A}}(T_E(X), F)$$
$$\xrightarrow{cr_E(X)_*} \mathrm{Tor}_n^{\mathcal{A}}(X, F) \to \mathrm{Tor}_{n-1}^{\mathcal{A}}(\kappa_E(X), F) \to \cdots$$

Mais le diagramme commutatif

$$\begin{array}{ccc} \mathrm{Tor}_n^{\mathcal{A}}(T_E(X), F) & \xrightarrow{cr_E(X)_*} & \mathrm{Tor}_n^{\mathcal{A}}(X, F) \\ \simeq \downarrow & \nearrow {\scriptstyle cr_E(F)_* = 0} & \\ \mathrm{Tor}_n^{\mathcal{A}}(X, T_E(F)) & & \end{array}$$

(F est polynomial de degré au plus d) montre que cette suite exacte longue se réduit à la nullité de $\mathrm{Tor}_0^{\mathcal{A}}(X, F)$ et des suites exactes courtes

$$0 \to \mathrm{Tor}_n^{\mathcal{A}}(X, F) \to \mathrm{Tor}_{n-1}^{\mathcal{A}}(\kappa_E(X), F) \to \mathrm{Tor}_{n-1}^{\mathcal{A}}(T_E(X), F) \to 0$$

pour $n > 0$. L'injectivité de la première flèche et l'hypothèse de récurrence appliquée à $\kappa_E(X)$ permettent de conclure. □

3.3. Le critère de Scorichenko

Le critère précédent s'avère très peu commode à vérifier en pratique : la surjectivité d'un effet croisé ne pose généralement pas trop de problème, mais la description du noyau devient souvent difficile, et celle des noyaux itérés plutôt désespérée. L'idée, remarquablement simple et efficace, de Scorichenko consiste à trouver une condition plus forte que la surjectivité des effets croisés, mais qui reste raisonnablement vérifiable dans de nombreux cas intéressants, et qui soit préservée par l'application du foncteur κ_E pour des raisons formelles.

Un premier critère simple, qu'on mentionne pour mémoire, est très naturel mais impose une condition un peu trop forte pour donner de nombreuses applications utiles :

Proposition 3.5. *Soient $d \in \mathbb{N}$, E un ensemble fini pointé de cardinal $d+2$, $F \in \operatorname{Ob} \mathcal{A}$-**Mod** polynomial de degré au plus d et $X \in \operatorname{Ob} \mathbf{Mod}\text{-}\mathcal{A}$ tel que l'effet croisé inverse $cr_E(X)$ soit un épimorphisme scindé. Alors $\operatorname{Tor}_*^{\mathcal{A}}(X, F) = 0$.*

Théorème 3.6 (Scorichenko). *Notons $\mathbf{M}(\mathcal{A})$ la sous-catégorie de $\mathcal{A}$ ayant les mêmes objets et dont les morphismes sont les monomorphismes scindés de $\mathcal{A}$ et θ : $\mathbf{Mod}\text{-}\mathcal{A} \to \mathbf{Mod}\text{-}\mathbf{M}(\mathcal{A})$ le foncteur de restriction à $\mathbf{M}(\mathcal{A})$. Soient $d \in \mathbb{N}$, E un ensemble fini pointé de cardinal $d+2$, $F \in \operatorname{Ob} \mathcal{A}$-**Mod** polynomial de degré au plus d et $X \in \operatorname{Ob} \mathbf{Mod}\text{-}\mathcal{A}$ tel que $\theta cr_E(X) : \theta T_E(X) \to \theta X$ soit un épimorphisme scindé. Alors $\operatorname{Tor}_*^{\mathcal{A}}(X, F) = 0$.*

Démonstration. Comme le foncteur θ est exact et fidèle, l'hypothèse faite sur X implique que c'est un objet de $\mathcal{K}_d$. Il suffit donc de vérifier qu'elle implique la même propriété pour $\kappa_E(X)$.

Pour cela, on note d'abord que la transformation naturelle $cr_E^{\mathcal{A}^{\mathrm{op}},\mathrm{inv}}$ s'obtient à partir des p_I de $\mathcal{A}^{\mathrm{op}}$, c'est-à-dire des u_I de $\mathcal{A}$, qui sont des monomorphismes scindés lorsque I est une partie non vide de E. On remarque par ailleurs que $T_E^2 := T_E \circ T_E \simeq T_{E\times E}$ (qu'on peut voir comme endofoncteur de $\mathbf{Mod}\text{-}\mathcal{A}$ ou $\mathbf{Mod}\text{-}\mathbf{M}(\mathcal{A})$) est muni d'une involution γ qui intervertit les deux facteurs du produit cartésien $E \times E$. Elle bénéficie de la propriété suivante : le diagramme

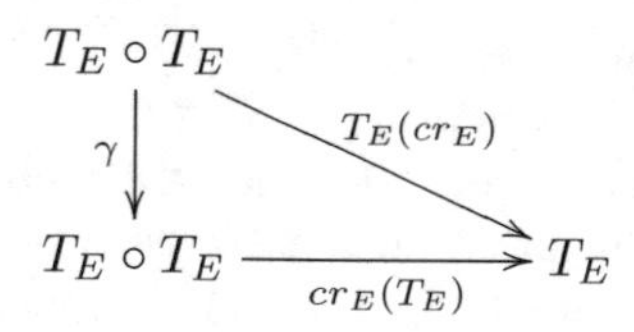

commute.

Considérons maintenant une section $s : \theta X \to \theta T_E(X)$ de $\theta cr_E(X)$. Notons s' le morphisme :

$$\theta T_E(X) = T_E(\theta X) \xrightarrow{T_E(s)} T_E(\theta T_E(X)) = \theta(T_E \circ T_E)(X) \xrightarrow{\gamma} \theta(T_E \circ T_E)(X).$$

C'est une section de $\theta cr_E(T_E(X))$ en raison de la commutativité du diagramme

$$\begin{array}{ccccccc}
T_E\theta X & \xrightarrow{T_E(s)} & T_E\theta T_E X & \xrightarrow{=} & \theta T_E^2 X & \xrightarrow{\theta\gamma_X} & \theta T_E^2 X \\
& \searrow^{Id} & \downarrow{\scriptstyle T_E\theta cr_E(X)} & & \downarrow{\scriptstyle \theta T_E cr_E(X)} & \swarrow_{\theta cr_E(T_E X)} & \\
& & T_E\theta X & \xrightarrow{=} & \theta T_E X & &
\end{array}$$

De plus, cette section est compatible à s et au morphisme $cr_E(X) : T_E(X) \to X$ en ce sens que le diagramme

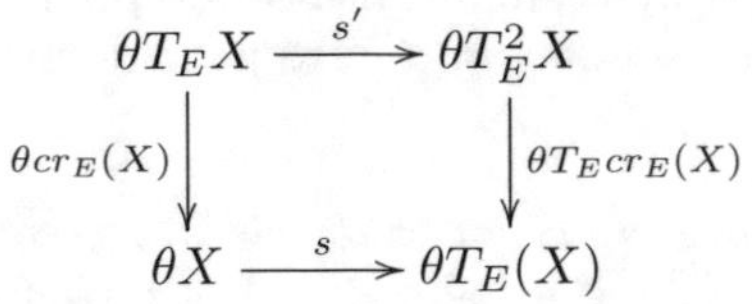

$$\begin{tikzcd} \theta T_E X \arrow[r, "s'"] \arrow[d, "\theta cr_E(X)"'] & \theta T_E^2 X \arrow[d, "\theta T_E cr_E(X)"] \\ \theta X \arrow[r, "s"] & \theta T_E(X) \end{tikzcd}$$

commute, parce que le diagramme

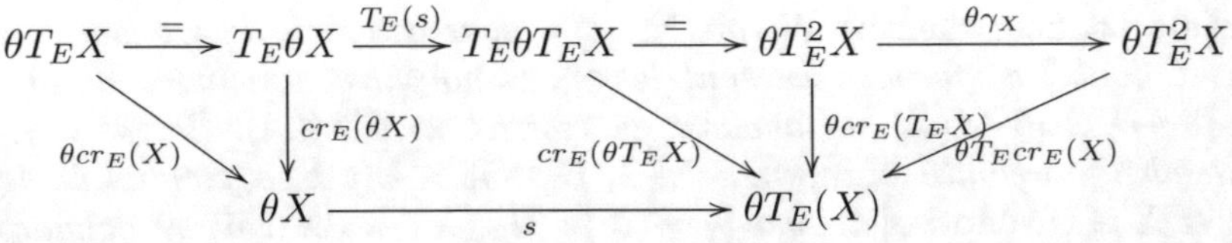

$$\begin{tikzcd} \theta T_E X \arrow[r, "="] \arrow[dr, "\theta cr_E(X)"'] & T_E \theta X \arrow[r, "T_E(s)"] \arrow[d, "cr_E(\theta X)"] & T_E \theta T_E X \arrow[r, "="] \arrow[dr, "cr_E(\theta T_E X)"'] & \theta T_E^2 X \arrow[r, "\theta \gamma_X"] \arrow[d, "\theta cr_E(T_E X)"] & \theta T_E^2 X \arrow[dl, "\theta T_E cr_E(X)"] \\ & \theta X \arrow[rr, "s"'] & & \theta T_E(X) & \end{tikzcd}$$

commute. Cela montre que s' induit une section de $cr_E(\kappa_E(X))$ et achève la démonstration. □

Corollaire 3.7. *Soient* $F \in \mathrm{Ob}\,\mathcal{A}$-**Mod** *analytique et* $X \in \mathrm{Ob}\,$**Mod**-$\mathcal{A}$ *tel que* $\theta cr_E(X) : \theta T_E(X) \to \theta X$ *soit un épimorphisme scindé pour tout ensemble fini pointé* E*. Alors* $\mathrm{Tor}_*^{\mathcal{A}}(X, F) = 0$.

Exemple 3.8. Supposons que $T : \mathcal{A} \to \mathbf{Ab}$ est un foncteur additif.

Alors $\mathrm{Tor}_*^{\mathcal{A}}(\bar{\Bbbk}^T, F) = 0$ (si E est un groupe abélien, on note $\bar{\Bbbk}^E$ le noyau de l'évaluation $\Bbbk^E \to \Bbbk$ en 0 ; cela définit un foncteur contravariant des groupes abéliens vers les $\Bbbk$-modules) si $F \in \mathrm{Ob}\,\mathcal{A}$-**Mod** est analytique.

Soit en effet E un ensemble fini pointé, de point de base a. Pour un objet A de $\mathcal{A}$, on définit une application linéaire $\bar{\Bbbk}^{T(A)} \to \bar{\Bbbk}^{T(A^{\oplus E})}$ en associant à une fonction $f : T(A) \to \Bbbk$ (nulle en 0) la fonction $g : T(A^{\oplus E}) \to \Bbbk$ définie par $g(y) = f(x)$ si $y = T(i)(x)$, où $i : A \to A^{\oplus E}$ est l'inclusion du facteur correspondant au point de base, $g(y) = 0$ sinon. On laisse au lecteur le soin de vérifier que cela définit une section de l'effet croisé fonctorielle en A relativement à $\mathbf{M}(\mathcal{A})$, permettant d'appliquer le corollaire précédent.

4. Deuxième description de l'homologie stable des groupes linéaires et unitaires à coefficients polynomiaux par l'homologie des foncteurs

Abstract. *We prove here the main result of this series. It relates stable homology of linear or unitary groups over a ring* A*, with coefficients twisted by a polynomial functor, to torsion groups on the category of finitely generated projective* A*-modules. We start from the second lecture's result and a suitable derived Kan extension, and we study the corresponding Grothendieck spectral sequence. Its collapsing is proved by the previous lecture's Scorichenko's criterion. It does not* *require the computation of the derived Kan extension (which seems out of reach), but only its qualitative properties as they result from the monoidal structures.*

Soient A un anneau muni d'une anti-involution, $\epsilon = 1$ ou -1, $\mathbf{P}(A)$ la catégorie des A-modules projectifs de type fini, $\mathbf{H}_\epsilon^{\deg}(A)$ la catégorie des A-modules projectifs de type fini munis d'une forme ϵ-hermitienne éventuellement dégénérée, et $\mathbf{H}_\epsilon(A)$ la sous-catégorie pleine des espaces non dégénérés.

Chacune de ces catégories est intéressante du point homologique pour les raisons suivantes.

Tout d'abord, $\mathbf{H}_\epsilon^{\deg}(A)$ se prête relativement bien aux calculs homologiques (au moins lorsque 2 est inversible dans A) car c'est la catégorie des objets de $\mathbf{P}(A)$ munis d'un élément de $\mathrm{S}^2_\epsilon(M)$, où S^2_ϵ est vu ici comme un foncteur contravariant ensembliste depuis $\mathbf{P}(A)$. Cela entraîne formellement la propriété qui suit :

Proposition 4.1. *Il existe un isomorphisme naturel*

$$\mathrm{Tor}_*^{\mathbf{H}_\epsilon^{\deg}(A)}(X, \iota(F)) \simeq \mathrm{Tor}_*^{\mathbf{P}(A)}(\omega(X), F)$$

de $\Bbbk$-modules gradués, où $\iota : \mathcal{A}\text{-}\mathbf{Mod} \to \mathbf{H}_\epsilon^{\deg}(A)\text{-}\mathbf{Mod}$ est le foncteur exact de précomposition par l'oubli de la forme hermitienne et le foncteur exact ω : $\mathbf{Mod}\text{-}\mathbf{H}_\epsilon^{\deg}(A) \to \mathbf{Mod}\text{-}\mathbf{P}(A)$ est donné sur les objets par

$$\omega(X)(M) = \bigoplus_{q \in \mathrm{S}^2_\epsilon(M)} X(M, q).$$

Corollaire 4.2. *Il existe un isomorphisme naturel*

$$H_*(\mathbf{H}_\epsilon^{\deg}(A), \iota(F)) \simeq \mathrm{Tor}_*^{\mathbf{P}(A)}(\Bbbk[\mathrm{S}^2_\epsilon], F)$$

de $\Bbbk$-modules gradués.

(Nous verrons dans la dernière partie de ce cours comment calculer de tels $\Bbbk$-modules dans les cas favorables.)

D'un autre côté, le résultat principal de la deuxième séance nous apprend que :

Proposition 4.3. *Pour tout foncteur $F \in \mathrm{Ob}\, \mathbf{H}_\epsilon(A)\text{-}\mathbf{Mod}$, il existe un isomorphisme naturel*

$$H_*(U_\infty^\epsilon(A); F_\infty) \simeq H_*(U_\infty^\epsilon(A) \times \mathbf{H}_\epsilon(A); F)$$

(avec action triviale de $U_\infty^\epsilon(A)$ à droite) de $\Bbbk$-modules gradués.

Dans cette séance, nous allons montrer que, lorsque F est l'image par le foncteur ι d'un foncteur analytique de $\mathcal{A}$-$\mathbf{Mod}$, les homologies de $\mathbf{H}_\epsilon(A)$ et $\mathbf{H}_\epsilon^{\deg}(A)$ à coefficients dans F coïncident.

4.1. Extensions de Kan dérivées

Soit $i : \mathcal{C} \to \mathcal{D}$ un foncteur entre petites catégories. Il existe une «adjonction»

$$i_!(G) \underset{\mathcal{D}}{\otimes} F \simeq G \underset{\mathcal{C}}{\otimes} i^*(F)$$

naturelle en $G \in \operatorname{Ob} \mathbf{Mod}\text{-}\mathcal{C}$ et $F \in \operatorname{Ob} \mathcal{D}\text{-}\mathbf{Mod}$, où

$$i_!(G)(D) := G \underset{\mathcal{C}}{\otimes} i^*(P^{\mathcal{D}}_D).$$

En général, le foncteur $i_!$ n'est pas exact, de sorte que cette adjonction ne se propage pas directement aux groupes de torsion supérieurs. Néanmoins, on peut dériver cette adjonction en une suite spectrale de Grothendieck (ou de foncteurs composés) – cf. par exemple [Wei94] Corollaire 5.8.4 (tout marche bien ici car $i_!$ préserve les projectifs) qui prend la forme

$$E^2_{p,q} = \operatorname{Tor}^{\mathcal{D}}_p(\mathbb{L}_q(i_!)(G), F) \Rightarrow \operatorname{Tor}^{\mathcal{C}}_{p+q}(G, i^*(F))$$

(avec une différentielle d^r sur E^r de degré $(-r, r-1)$), où les $\mathbb{L}_q(i_!)$ sont les foncteurs dérivés à gauche du foncteur exact à droite $i_!$, qui sont donnés par

$$\mathbb{L}_q(i_!)(G)(D) = \operatorname{Tor}^{\mathcal{C}}_q(G, i^*(P^{\mathcal{D}}_D)).$$

En particulier, prenant pour G le foncteur constant $\Bbbk$, on obtient une suite spectrale

$$E^2_{p,q} = \operatorname{Tor}^{\mathcal{D}}_p(L_q, F) \Rightarrow H_{p+q}(\mathcal{C}; i^*(F))$$

où

$$L_q(D) = H_q(\mathcal{C}/D; \Bbbk)$$

où $\mathcal{C}/D$ désigne la catégorie des objets C de $\mathcal{C}$ munis d'un morphisme $D \to i(C)$ dans $\mathcal{D}$, les flèches étant celles de $\mathcal{C}$ qui font commuter le diagramme évident (l'équivalence avec la formule précédente s'obtient par un jeu facile d'adjonction). Si l'on note $X_q(D) = \tilde{H}_q(\mathcal{C}/D; \Bbbk)$, on a donc la proposition suivante :

Proposition 4.4. *Si $F \in \operatorname{Ob} \mathcal{D}\text{-}\mathbf{Mod}$ vérifie*

$$\operatorname{Tor}^{\mathcal{D}}_*(X_q, F) = 0$$

pour tout $q \in \mathbb{N}$, alors le foncteur i induit un isomorphisme

$$i_* : H_*(\mathcal{C}; i^*(F)) \xrightarrow{\simeq} H_*(\mathcal{D}; F)$$

en homologie à coefficients dans F.

Nous allons appliquer cette propriété lorsque $i : \mathbf{H}_\epsilon(A) \hookrightarrow \mathbf{H}^{\deg}_\epsilon(A)$ est le foncteur d'inclusion et F l'image par le foncteur ι d'un foncteur analytique de $\mathbf{P}(A)$-$\mathbf{Mod}$. Par la propriété 4.1, on voit qu'on peut le faire en appliquant le critère de Scorichenko (Corollaire 3.7) aux foncteurs $\omega(X_q)$.

4.2. Cas de l'inclusion $\mathbf{H}_\epsilon(A) \hookrightarrow \mathbf{H}_\epsilon^{\deg}(A)$: préparation

Il n'est pas question de calculer les valeurs des foncteurs X_q. Nous utiliserons uniquement les propriétés suivantes :

1. X_q envoie l'objet initial 0 de $\mathbf{H}_\epsilon^{\deg}(A)$ sur 0 ;
2. X_q envoie toute inclusion canonique $M \hookrightarrow M \overset{\perp}{\oplus} H$, où M est un objet de $\mathbf{H}_\epsilon^{\deg}(A)$ et H un objet de $\mathbf{H}_\epsilon(A)$, sur un isomorphisme.

La première propriété résulte de ce que $\mathbf{H}_\epsilon(A)/0$ a un objet initial. Pour la seconde, on note que le foncteur $\mathbf{H}_\epsilon(A)/M \to \mathbf{H}_\epsilon(A)/(M \overset{\perp}{\oplus} H)$ induit par $-\overset{\perp}{\oplus}H$ est une équivalence de catégories. Cela provient essentiellement de l'existence d'un supplémentaire orthogonal pour tout sous-espace *non dégénéré* d'un espace hermitien : un quasi-inverse de ce foncteur est donné en associant à $(N, f : M \overset{\perp}{\oplus} H \to N)$ l'objet $(f(H)^\perp (\subset N), \bar{f} : M \to f(H)^\perp)$ de $\mathbf{H}_\epsilon(A)/M$.

Lemme 4.5. *Soit V un A-module projectif de type fini. Il existe un objet $H(V)$ de $\mathbf{H}_\epsilon(A)$ vérifiant, pour toute forme hermitienne (éventuellement dégénérée) q sur V, les propriétés suivantes :*

1. *il existe un morphisme $\alpha : (V, q) \to H(V)$ de $\mathbf{H}_\epsilon^{\deg}(A)$ dont l'application A-linéaire sous-jacente appartient à $\mathbf{M}(A)$;*
2. *de plus, pour $q = 0$, on peut choisir α de sorte que tout morphisme de $(V, 0)$ vers un objet de $\mathbf{H}_\epsilon(A)$ dont le morphisme sous-jacent de $\mathbf{P}(A)$ appartienne à $\mathbf{M}(A)$ se factorise par α ;*
3. *pour tous morphismes $f : V \to W$ de $\mathbf{M}(A)$ et $\varphi : (W, 0) \to E$ de $\mathbf{H}_\epsilon^{\deg}(A)$, avec E non dégénéré, il existe une factorisation*

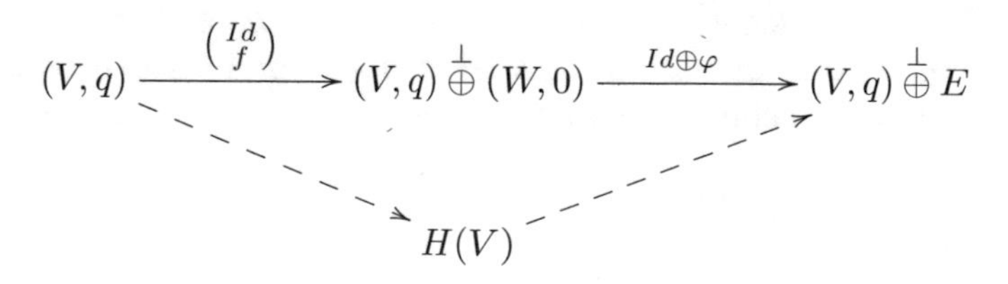

Nous ne donnerons pas les détails, élémentaires et classiques, de cette propriété. Bornons-nous à mentionner que l'on prend pour $H(V)$ le A-module $V \oplus V^*$ muni de la forme hermitienne de matrice [7] $\begin{pmatrix} 0 & 1_{M^*} \\ \epsilon.1_M & 0 \end{pmatrix}$. Une bonne référence pour les espaces hermitiens sur un anneau quelconque est [Knu91] ; on trouvera aussi une démonstration complète du lemme précédent dans [Dja12].

---

7. Plus précisément, il s'agit de la matrice de la forme sesquilinéaire associée $M \oplus M^* \to (M \oplus M^*)^* \simeq M^* \oplus M$ – le morphisme canonique $M \to (M^*)^*$ est un isomorphisme car M est projectif de type fini.

4.3. Le résultat principal

On note ici ι' la précomposition par l'oubli de la forme hermitienne $\mathbf{H}_\epsilon(A) \to \mathbf{P}(A)$.

Théorème 4.6 ([DV10] pour A corps fini, [Dja12] pour le cas général). *Si $F \in \operatorname{Ob}\mathbf{P}(A)$-**Mod** est analytique, l'inclusion $\mathbf{H}_\epsilon(A) \hookrightarrow \mathbf{H}_\epsilon^{\deg}(A)$ induit un isomorphisme*

$$H_*(\mathbf{H}_\epsilon(A); \iota'(F)) \xrightarrow{\simeq} H_*(\mathbf{H}_\epsilon^{\deg}(A); \iota(F)).$$

Démonstration. D'après ce qui précède, il suffit de construire des sections faiblement fonctorielles aux applications $\Bbbk$-linéaires

$$\begin{aligned} cr_E(\omega(X_n))(V) : \omega(X_n)(V^{\oplus E}) &= \bigoplus_{Q \in \mathrm{S}^2_\epsilon(V^{\oplus E})} X_n(V^{\oplus E}, Q) \to \omega(X_n)(V) \\ &= \bigoplus_{q \in \mathrm{S}^2_\epsilon(V)} X_n(V, q) \end{aligned}$$

où E est un ensemble pointé fini (de point de base noté e), n un entier naturel et V un A-module projectif de type fini ; ici faiblement fonctoriel signifie fonctoriel relativement aux monomorphismes scindés. Notons $\pi = p_e : V^{\oplus E} \to V$ la projection sur le facteur correspondant à e et définissons une application linéaire $s(V) : \omega(X_n)(V) \to \omega(X_n)(V^{\oplus E})$ par sa composante sur $X_n(V, q)$ donnée par

$$\begin{aligned} X_n(V, q) \simeq X_n\big((V, q) \overset{\perp}{\oplus} H(V^{\oplus E\setminus e})\big) &\xrightarrow{X_n\left((V,q)\overset{\perp}{\oplus}\alpha\right)} X_n\big((V, q) \overset{\perp}{\oplus} (V^{\oplus E\setminus e}, 0)\big) \\ &= \cdots X_n(V^{\oplus E}, \pi^* q) \hookrightarrow \omega(X_n)(V^{\oplus E}) \end{aligned}$$

où $\alpha : (V^{\oplus E\setminus e}, 0) \to H(V^{\oplus E\setminus e})$ est comme dans le Lemme 4.5. Vérifions d'abord que cette définition ne dépend pas du choix de α. Supposons $\beta : (V, 0) \to L$ donné, avec $L \in \operatorname{Ob}\mathbf{H}_\epsilon(A)$ et le morphisme de $\mathbf{P}(A)$ sous-jacent à β appartenant à $\mathbf{M}(A)$. Alors il existe (par le lemme) une factorisation

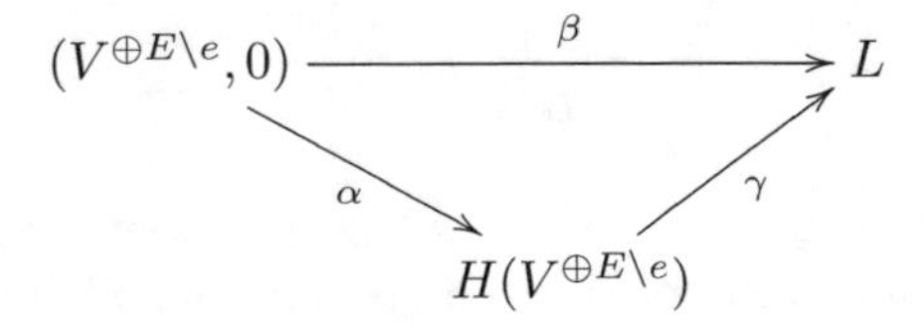

ce qui implique que les diagrammes

$$\begin{array}{ccccc} (V,q)\overset{\perp}{\oplus}(V^{\oplus E\setminus e},0) & \xrightarrow{(V,q)\overset{\perp}{\oplus}\beta} & (V,q)\overset{\perp}{\oplus}L & \longleftarrow & (V,q) \\ & \searrow^{(V,q)\overset{\perp}{\oplus}\alpha} & \nearrow_{(V,q)\overset{\perp}{\oplus}\gamma} & \swarrow & \\ & & (V,q)\overset{\perp}{\oplus}H(V^{\oplus E\setminus e}) & & \end{array}$$

(où les deux flèches non spécifiées sont les inclusions canoniques) puis

$$\begin{array}{ccccc} X_n(V,q) & \xrightarrow{\simeq} & X_n\big((V,q)\overset{\perp}{\oplus} H(V^{\oplus E\setminus e})\big) & \xrightarrow{X_n\left((V,q)\overset{\perp}{\oplus}\alpha\right)} & X_n\big((V,q)\overset{\perp}{\oplus}(V^{\oplus E\setminus e},0)\big) \\ \uparrow{=} & & \uparrow{X_n\left((V,q)\overset{\perp}{\oplus}\gamma\right)} & & \uparrow{=} \\ X_n(V,q) & \xrightarrow{\simeq} & X_n\big((V,q)\overset{\perp}{\oplus} L\big) & \xrightarrow{X_n\left((V,q)\overset{\perp}{\oplus}\beta\right)} & X_n\big((V,q)\overset{\perp}{\oplus}(V^{\oplus E\setminus e},0)\big) \end{array}$$

commutent.

Cela établit non seulement l'indépendance de $s(V)$ envers le choix de α, mais aussi le caractère faiblement fonctoriel de cette application : si $u : V \to W$ est un morphisme de $\mathbf{M}(A)$, on peut appliquer ce qui précède à

$$(V^{\oplus E\setminus e},0) \xrightarrow{u} (W^{\oplus E\setminus e},0) \xrightarrow{\alpha_W} H(W^{\oplus E\setminus e},0)$$

qui vérifie toutes les hypothèses requises pour β.

Examinons maintenant les composées

$$\omega(X_n)(V) \xrightarrow{s(V)} \omega(X_n)(V^{\oplus E\setminus e}) \xrightarrow{u_I^*} \omega(X_n)(V)$$

où I est une partie pointée de E. Leurs composantes sont données par

$$X_n(V,q) \simeq X_n\big((V,q)\overset{\perp}{\oplus} H(V^{\oplus E\setminus e})\big) \xrightarrow{X_n\left((V,q)\overset{\perp}{\oplus}\alpha\right)} X_n\big((V,q)\overset{\perp}{\oplus}(V^{\oplus E\setminus e},0)\big)$$
$$= \cdots X_n(V^{\oplus E},\pi^* q) \xrightarrow{u_I^*} X_n(V, u_I^*\pi^* q) = X_n(V,q)$$

(on $\pi u_I = Id_V$ car $e \in I$). Pour I réduit à e, c'est l'identité, car $u_e : (V,q) \to (V,q)\overset{\perp}{\oplus}(V^{\oplus E\setminus e},0)$ est l'inclusion canonique.

Si I contient strictement e, alors on peut trouver un diagramme commutatif

$$\begin{array}{ccccc} (V,q) & \xrightarrow{u_I=\begin{pmatrix} Id \\ u_J \end{pmatrix}} & (V,q)\overset{\perp}{\oplus}(V^{\oplus E\setminus e},0) & \xrightarrow{(V,q)\overset{\perp}{\oplus}\alpha} & (V,q)\overset{\perp}{\oplus} H(V^{\oplus E\setminus e}) \\ & \searrow & & \nearrow & \\ & & H(V^{\oplus E\setminus e}) & & \end{array}$$

grâce au Lemme 4.5, où $J = I\setminus e$ (qui est non vide, donc u_J est un monomorphisme scindé). Comme $H(V^{\oplus E\setminus e})$ est non dégénéré, X_n l'envoie sur 0, et notre composée est donc nulle. Finalement, $s(V)$ est bien (au signe près) une section de l'effet croisé $cr_E(\omega(X_n))(V)$, d'où le théorème. □

Cas particulier : le théorème de Scorichenko sur l'homologie des groupes linéaires. Nous donnons maintenant le résultat fondamental de [Sco00] (avec nos notations), qui constitue un cas particulier du théorème 4.6.

Théorème 4.7 (Scorichenko). *Soit* $B : \mathbf{P}(A)^{\mathrm{op}} \times \mathbf{P}(A) \to \Bbbk\text{-}\mathbf{Mod}$ *un bifoncteur polynomial. Il existe un isomorphisme naturel*

$$H_*(GL_\infty(A); B_\infty) \simeq HH_*(GL_\infty(A) \times \mathbf{P}(A); B)$$

de $\Bbbk$*-modules gradués.*

Dans le cas où $A = \Bbbk$ est un corps fini, l'homologie de $GL_\infty(\Bbbk)$ à coefficients dans $\Bbbk$ est triviale (Quillen [Qui72]), on peut donc se débarrasser du terme GL_∞ dans le membre de droite.

On déduit le théorème précédent du théorème 4.6 en considérant l'anneau $A^{\mathrm{op}} \times A$ muni de l'anti-involution échangeant les deux facteurs : ses groupes unitaires ne sont autres que les groupes linéaires sur A, et obtenir l'homologie de Hochschild à partir de l'homologie de la catégorie des $A^{\mathrm{op}} \times A$-espaces hermitiens (éventuellement dégénérés ; ici $\epsilon = 1$) est un jeu élémentaire d'adjonction (à l'aide de la *catégorie des factorisations* de Quillen, introduite dans [Qui73]).

5. Exemples de calculs et autres applications

Abstract. *The last lecture of the series explains how to carry out computations in good cases, and how to prove qualitative properties of the functor homology groups that appear in the main theorem of the previous lecture. The lecture illustrates classical computational tools in functor homology, such as exponential functors. Qualitative properties such as base change are obtained by simple adjunction arguments.*

On va expliquer comment mener certains calculs de $\Bbbk$-modules gradués du type

$$\mathrm{Tor}_*^{\mathbf{P}(A)}(\Bbbk[S_A^2]^\vee, F)$$

pour un foncteur polynomial F, où A est un anneau commutatif muni de l'involution triviale (et $\epsilon = 1$, i.e., on traite de groupes orthogonaux plutôt que de groupes symplectiques) – cette restriction, destinée surtout à alléger les notations, est contingente – et dans lequel 2 est inversible – cette hypothèse s'avère en revanche fondamentale pour que les calculs ne soient pas inextricables (dans [DV10], on donne toutefois quelques résultats très partiels en caractéristique 2).

5.1. Réductions du problème

On commence par se ramener à calculer

$$\mathrm{Tor}_*^{\mathbf{P}(A)}(\Bbbk[T_A^2]^\vee, F)$$

dont le précédent groupe de torsion est facteur direct grâce à l'inversibilité de 2 dans A (qui garantit que S_A^2 est facteur direct dans T_A^2). Cette étape n'est pas innocente : même quand on sait déterminer entièrement ces groupes de torsion, la détermination explicite de l'idempotent donnant le facteur qui nous intéresse constitue tout sauf un problème trivial. Néanmoins, on peut y arriver dans certains cas.

Le $\Bbbk$-module gradué précédent n'est autre qu'une homologie de Hochschild :

$$\begin{aligned}\mathrm{Tor}_*^{\mathbf{P}(A)}(\Bbbk[T_A^2]^\vee, F) &\simeq \mathrm{Tor}_*^{\mathbf{P}(A)\times\mathbf{P}(A)}((M,N)\mapsto \Bbbk[(M\otimes N)^*], \oplus^* F)\\ &\simeq \mathrm{Tor}_*^{\mathbf{P}(A)^{\mathrm{op}}\times\mathbf{P}(A)}((U,V)\mapsto \Bbbk[\mathrm{Hom}_A(V,U)], s^*F)\\ &\simeq HH_*(\mathbf{P}(A); s^*F)\end{aligned}$$

où $s : \mathbf{P}(A)^{\mathrm{op}} \times \mathbf{P}(A) \to \mathbf{P}(A) \quad (U,V) \mapsto U^* \oplus V$, où le premier isomorphisme repose sur l'adjonction entre diagonale et somme directe et la seconde sur l'isomorphisme canonique $(U^* \otimes V)^* \simeq \mathrm{Hom}_A(V,U)$ pour U,V projectifs de type fini.

Nous nous concentrerons maintenant sur quelques cas particuliers importants :

1. F est (un terme) d'un foncteur exponentiel gradué de Hopf $E^\bullet$, i.e., d'une suite de foncteurs $(E^n)_{n\in\mathbb{N}}$ telle que $E^0 = \Bbbk$ et munie d'isomorphismes naturels
$$E^n(U\oplus V) \simeq \bigoplus_{i+j=n} E^i(U)\otimes E^j(V)$$
(les produits tensoriels sont implicitement pris sur $\Bbbk$) vérifiant des conditions de compatibilité (une référence pour cette notion est [FFSS99], par exemple, ou l'appendice B de [DV10]), notamment une propriété d'ϵ-commutativité (où $\epsilon \in \{-1,1\}$ vaut 1 pour les puissances symétriques ou divisées et -1 pour les puissances extérieures) ;
2. F est une puissance tensorielle sur $\Bbbk$ (on suppose alors que A est une $\Bbbk$-algèbre) ; on bénéficie alors (« formule du binôme ») d'isomorphismes naturels
$$T^n_\Bbbk(U\oplus V) \simeq \bigoplus_{i+j=n} (T^i_\Bbbk(U)\otimes T^j_\Bbbk(V)) \uparrow^{\Sigma_n}_{\Sigma_i\times\Sigma_j}$$
vérifiant encore des conditions de compatibilité.

On peut donc ramener le calcul de $\mathrm{Tor}_*^{\mathbf{P}(A)}(\Bbbk[T_A^2]^\vee, F)$ à celui de groupes de torsion entre foncteurs polynomiaux usuels sur $\mathbf{P}(A)$ (par exemple, dans le premier cas, à $\mathrm{Tor}_*((E^i)^\vee, E^j)$, en supposant que les E^i prennent des valeurs $\Bbbk$-plates). On dispose des renseignements suivants sur l'idempotent qui nous intéresse :

- l'involution qui échange les deux facteurs dans T_A^2 se lit sur les $\mathrm{Tor}_*((E^i)^\vee, E^j)$ (dans le premier cas) ou les $\mathrm{Tor}_*((T^i_\Bbbk)^\vee, T^j_\Bbbk)$ (dans le second) grâce à l'involution canonique de $\mathrm{Tor}_*(X^\vee, X)$ existant pour tout foncteur X (on dispose d'un isomorphisme d'adjonction $\mathrm{Tor}_*(Y^\vee, X) \simeq \mathrm{Tor}_*(X^\vee, Y)$ qui donne cette involution pour $Y = X$), au signe ϵ^{ij} près dans le premier cas ;
- on peut décrire l'idempotent qui nous intéresse – ou plus exactement, le morphisme induit par $x \mapsto x + \bar{x}$ sur T_A^2 (qui a la même image puisque 2 est inversible) – comme la convolution entre l'identité et l'involution précédente, où l'on utilise la structure de foncteur de Hopf (ou son analogue dans le second cas) pour munir les groupes de torsion qu'on veut calculer d'une structure d'algèbre de Hopf (bigraduée).

5.2. Quelques calculs

En se fondant sur les calculs puissants de [FFSS99] (qui procurent la structure d'algèbre de Hopf dont on a besoin et permettent de comprendre l'involution susmentionnée), on parvient par exemple au résultat suivant (qu'on a exprimé en cohomologie pour le rendre plus intuitif) :

Théorème 5.1 (Cf. [DV10]). *Soit k un corps fini de cardinal q impair. L'algèbre de cohomologie stable des groupes orthogonaux sur k à coefficients dans les puissances symétriques est polynomiale sur des générateurs de bidegré $(2q^s m, q^s + 1)$ (où le premier degré est le degré homologique et le second le degré interne) indexés par les entiers naturels m et s.*

Dans [Dja12], on montre le résultat suivant, qu'on peut obtenir à partir de la considération de la seconde puissance extérieure et du théorème de Hochschild–Kostant–Rosenberg, qui décrit (en particulier) l'homologie de Hochschild d'un corps commutatif.

Proposition 5.2. *Soient k un corps commutatif de caractéristique nulle ; pour $n \in \mathbb{N}$, notons $\mathfrak{o}_{n,n}(k)$ la représentation adjointe de $O_{n,n}(k)$. Alors le k-espace vectoriel gradué*

$$\operatorname*{colim}_{n\in\mathbb{N}} H_*(O_{n,n}(k); \mathfrak{o}_{n,n}(k))$$

est isomorphe au produit tensoriel (sur $\mathbb{Q}$) de $H_*(O_{\infty,\infty}(k); \mathbb{Q})$ (où $O_{\infty,\infty}$ est la colimite des $O_{n,n}$) et de la partie impaire de la k-algèbre graduée $\Omega_k^* = \Lambda^*(\Omega_k^1)$ des différentielles de Kähler de k (vu comme $\mathbb{Q}$-algèbre). En particulier, cet espace vectoriel gradué est nul si k est une extension algébrique de $\mathbb{Q}$.

On peut aussi mener certains calculs sur des anneaux qui ne sont pas des corps : dans [Dja12], on donne plusieurs énoncés pour des anneaux (commutatifs avec 2 inversible) sans torsion sur $\mathbb{Z}$.

Il est à noter (voir [Dja12]) que la détermination de $\mathrm{Tor}_*^{\mathbf{P}(A)}(\Bbbk[S_A^2]^\vee, F)$ comme $\Bbbk$-module gradué, lorsque F est une puissance tensorielle, dépend uniquement de la connaissance de $\mathrm{Tor}_*^{\mathbf{P}(A)}(Id^\vee, Id)$ comme module gradué : la connaissance de l'involution sur ces groupes de torsion n'est pas nécessaire.

5.3. Quelques propriétés qualitatives

Une observation élémentaire mais très efficace de Touzé (voir [Tou10], qui est déclinée dans [DV10] dans le cadre des groupes discrets qui est le nôtre) sur les produits en cohomologie des foncteurs permet de déduire de la comparaison entre (co)homologie des foncteurs et (co)homologie stable des groupes orthogonaux le résultat suivant :

Proposition 5.3. *Soient k un corps fini de cardinal impair, $\Bbbk = k$, F et G deux foncteurs polynomiaux de $\mathbf{P}(k)$-$\mathbf{Mod}$. Alors le produit externe*

$$H^*(O_{\infty,\infty}(k); F_\infty) \otimes H^*(O_{\infty,\infty}(k); G_\infty) \to H^*(O_{\infty,\infty}(k); F_\infty \otimes G_\infty)$$

est un monomorphisme naturellement scindé.

Proposition 5.4. *Supposons que k est une extension finie (ou même seulement algébrique) de $\mathbb{Q}$, que $\Bbbk = \mathbb{Q}$ et que $F \in \operatorname{Ob} \mathbf{P}(k)\text{-}\mathbf{Mod}$ est un foncteur polynomial. Alors il existe un isomorphisme naturel*

$$H_*(O_{\infty,\infty}(k); F_\infty) \simeq H_*(O_{\infty,\infty}(k); \mathbb{Q}) \underset{\mathbb{Q}}{\otimes} H_0(O_{\infty,\infty}(k); F_\infty)$$

de $\mathbb{Q}$-espaces vectoriels gradués.

La démonstration repose sur le caractère semi-simple de la catégorie des foncteurs polynomiaux de $\mathbf{P}(k)$-**Mod** lorsque k est un corps de nombres et sur la trivialité des auto-extensions du foncteur identité des $\mathbb{Q}$-espaces vectoriels (qui s'appuie sur la résolution barre – cf. par exemple [FP98], qui traite du cas beaucoup plus difficile des entiers, dont on peut déduire cette annulation).

Remarque 5.5. Sous certaines hypothèses (par exemple, sur des corps), on peut obtenir des résultats analogues aux précédents avec des groupes orthogonaux non nécessairement hyperboliques – par exemple euclidiens. Cela est discuté dans la dernière partie de [Dja12] (le cas général reste toutefois mystérieux dans ce genre de situation).

Un autre aspect agréable de l'homologie des foncteurs réside dans les changements de base qui se prêtent souvent à des comparaisons assez simples (par des arguments d'adjonction). Voici un cas particulier spectaculaire, lié d'après le Théorème 4.7 de Scorichenko à l'homologie stable des groupes linéaires à coefficients dans la représentation adjointe.

Soient A un anneau commutatif et S une partie multiplicative de A ; choisissons $\Bbbk = \mathbb{Z}$ (ou A). Alors

$$HH_*(\mathbf{P}(A[S^{-1}]); \operatorname{Hom}_{A[S^{-1}]}) \simeq HH_*(\mathbf{P}(A); \operatorname{Hom}_A)[S^{-1}].$$

On note en revanche qu'il est extrêmement difficile de comparer les homologies à coefficients constants de $GL_\infty(A[S^{-1}])$ et $GL_\infty(A)$.

Références

[AM04] Alejandro Adem and R. James Milgram. *Cohomology of finite groups*, volume 309 of *Grundlehren der Mathematischen Wissenschaften [Fundamental Principles of Mathematical Sciences]*. Springer-Verlag, Berlin, second edition, 2004.

[Bet89] Stanislaw Betley. Vanishing theorems for homology of $\mathrm{Gl}_n R$. *J. Pure Appl. Algebra*, 58(3):213–226, 1989.

[Bet99] Stanislaw Betley. Stable K-theory of finite fields. *K-Theory*, 17(2):103–111, 1999.

[Bet02] Stanislaw Betley. Twisted homology of symmetric groups. *Proc. Amer. Math. Soc.*, 130(12):3439–3445 (electronic), 2002.

[Cha84] Ruth Charney. On the problem of homology stability for congruence subgroups. *Comm. Algebra*, 12(17–18):2081–2123, 1984.

[Col11] G. Collinet. Homology stability for unitary groups over S-arithmetic rings. *J. K-Theory*, 8(2):293–322, 2011.

[CPSK77] E. Cline, B. Parshall, L. Scott, and Wilberd van der Kallen. Rational and generic cohomology. *Invent. Math.*, 39(2):143–163, 1977.

[Dja12] Aurélien Djament. Sur l'homologie des groupes unitaires à coefficients polynomiaux. *J. K-Theory*, 10(1):87–139, 2012.

[Dup01] Johan L. Dupont. *Scissors congruences, group homology and characteristic classes*, volume 1 of *Nankai Tracts in Mathematics*. World Scientific Publishing Co. Inc., River Edge, NJ, 2001.

[DV10] Aurélien Djament and Christine Vespa. Sur l'homologie des groupes orthogonaux et symplectiques à coefficients tordus. *Ann. Sci. Éc. Norm. Supér. (4)*, 43(3):395–459, 2010.

[Dwy80] W.G. Dwyer. Twisted homological stability for general linear groups. *Ann. of Math. (2)*, 111(2):239–251, 1980.

[EML54] Samuel Eilenberg and Saunders Mac Lane. On the groups $H(\Pi, n)$. II. Methods of computation. *Ann. of Math. (2)*, 60:49–139, 1954.

[FFSS99] Vincent Franjou, Eric M. Friedlander, Alexander Scorichenko, and Andrei Suslin. General linear and functor cohomology over finite fields. *Ann. of Math. (2)*, 150(2):663–728, 1999.

[FP78] Zbigniew Fiedorowicz and Stewart Priddy. *Homology of classical groups over finite fields and their associated infinite loop spaces*, volume 674 of *Lecture Notes in Mathematics*. Springer, Berlin, 1978.

[FP98] Vincent Franjou and Teimuraz Pirashvili. On the Mac Lane cohomology for the ring of integers. *Topology*, 37(1):109–114, 1998.

[Gal] Soren Galatius. Stable homology of automorphism groups of free groups. *Ann. of Math. (2)*, 173(2):705–768, 2011.

[Hat95] Allen Hatcher. Homological stability for automorphism groups of free groups. *Comment. Math. Helv.*, 70(1):39–62, 1995.

[Kar80] Max Karoubi. Théorie de Quillen et homologie du groupe orthogonal. *Ann. of Math. (2)*, 112(2):207–257, 1980.

[Kar83] Max Karoubi. Homology of the infinite orthogonal and symplectic groups over algebraically closed fields. *Invent. Math.*, 73(2):247–250, 1983. An appendix to the paper: "On the K-theory of algebraically closed fields" by A. Suslin.

[Knu91] Max-Albert Knus. *Quadratic and Hermitian forms over rings*, volume 294 of *Grundlehren der Mathematischen Wissenschaften [Fundamental Principles of Mathematical Sciences]*. Springer-Verlag, Berlin, 1991. With a foreword by I. Bertuccioni.

[Lod76] Jean-Louis Loday. K-théorie algébrique et représentations de groupes. *Ann. Sci. École Norm. Sup. (4)*, 9(3):309–377, 1976.

[Mil83] J. Milnor. On the homology of Lie groups made discrete. *Comment. Math. Helv.*, 58(1):72–85, 1983.

[ML98] Saunders Mac Lane. *Categories for the working mathematician*, volume 5 of *Graduate Texts in Mathematics*. Springer-Verlag, New York, second edition, 1998.

[MvdK02] B. Mirzaii and W. van der Kallen. Homology stability for unitary groups. *Doc. Math.*, 7:143–166 (electronic), 2002.

[Nak60] Minoru Nakaoka. Decomposition theorem for homology groups of symmetric groups. *Ann. of Math.* (2), 71:16–42, 1960.

[Pir00a] Teimuraz Pirashvili. Dold–Kan type theorem for Γ-groups. *Math. Ann.*, 318(2):277–298, 2000.

[Pir00b] Teimuraz Pirashvili. Hodge decomposition for higher order Hochschild homology. *Ann. Sci. École Norm. Sup.* (4), 33(2):151–179, 2000.

[Qui72] Daniel Quillen. On the cohomology and K-theory of the general linear groups over a finite field. *Ann. of Math.* (2), 96:552–586, 1972.

[Qui73] Daniel Quillen. Higher algebraic K-theory. I. In *Algebraic K-theory, I: Higher K-theories* (*Proc. Conf., Battelle Memorial Inst., Seattle, Wash.*, 1972), pages 85–147. Lecture Notes in Math., Vol. 341. Springer, Berlin, 1973.

[Sco00] Alexander Scorichenko. *Stable K-theory and functor homology over a ring.* PhD thesis, Evanston, 2000.

[Sus83] A. Suslin. On the K-theory of algebraically closed fields. *Invent. Math.*, 73(2):241–245, 1983.

[Sus87] A.A. Suslin. Algebraic K-theory of fields. In *Proceedings of the International Congress of Mathematicians, Vol.* 1, 2 (*Berkeley, Calif.*, 1986), pages 222–244, Providence, RI, 1987. Amer. Math. Soc.

[Tou10] Antoine Touzé. Cohomology of classical algebraic groups from the functorial viewpoint. *Adv. Math.*, 225(1):33–68, 2010.

[vdK80] Wilberd van der Kallen. Homology stability for linear groups. *Invent. Math.*, 60(3):269–295, 1980.

[Vog82] K. Vogtmann. A Stiefel complex for the orthogonal group of a field. *Comment. Math. Helv.*, 57(1):11–21, 1982.

[Wei94] Charles A. Weibel. *An introduction to homological algebra*, volume 38 of *Cambridge Studies in Advanced Mathematics*. Cambridge University Press, Cambridge, 1994.

Aurélien Djament
CNRS, Laboratoire de mathématiques Jean Leray (UMR 6629)
Faculté des Sciences
2, rue de la Houssinière
BP 92208
F-44322 Nantes cedex 3, France
e-mail: `Aurelien.Djament@univ-nantes.fr`

Progress in Mathematics, Vol. 311, 41–65

Lectures on Bifunctors and Finite Generation of Rational Cohomology Algebras

Wilberd van der Kallen

Abstract. This text is an updated version of material used for a course at Université de Nantes, part of 'Functor homology and applications', April 23–27, 2012. The proof [30], [31] by Touzé of my conjecture on cohomological finite generation (CFG) has been one of the successes of functor homology. We will not treat this proof in any detail. Instead we will focus on a formality conjecture of Chałupnik and discuss ingredients of a second generation proof [33] of the existence of the universal classes of Touzé.

Mathematics Subject Classification (2010). 20G10, 18G10.

Keywords. Strict polynomial functor, Schur algebra, cohomology algebra, finite generation, Frobenius twist, formality, reductive, divided power, derived category, collapsing spectral sequence.

1. The CFG theorem

In its most basic form the CFG theorem of [31] reads

Theorem 1.1 (Cohomological finite Generation). *Let G be a reductive algebraic group over an algebraically closed field k, and let A be a finitely generated commutative k-algebra on which G acts algebraically via algebra automorphisms. Then the cohomology algebra $H^*(G, A)$ is a finitely generated graded k-algebra.*

An essential ingredient in the proof of this theorem is the existence of certain universal cohomology classes. They were constructed by Touzé in [30]. We will discuss methods used in the new construction [33] of these classes.

2. Some history

Let us give some background. First there is *invariant theory* [3], [15], [28]. Classical invariant theory looked at the following situation. (We will give a very biased description, full of anachronisms.) Say the algebraic Lie group $G(\mathbb{C}) := SL_n(\mathbb{C})$

acts on a finite-dimensional complex vector space V with dual $V^\vee$. Then it also acts on the symmetric algebra $A = S^*_{\mathbb{C}}(V^\vee)$ of polynomial maps from V to $\mathbb{C}$. One is interested in the subalgebra $A^{G(\mathbb{C})}$ of elements fixed by $G(\mathbb{C})$. It is called the subalgebra of *invariants*. More generally, if W is another finite-dimensional complex vector space on which $G(\mathbb{C})$ acts, then $W \otimes_{\mathbb{C}} A$ encodes the polynomial maps from V to W. The subspace $(W \otimes_{\mathbb{C}} A)^G$ of fixed points or invariants in $W \otimes_{\mathbb{C}} A$ corresponds with the equivariant polynomial maps from V to W. This subspace of invariants is a module over the algebra of invariants $A^{G(\mathbb{C})}$. When $n = 2$ and V is irreducible Gordan (1868) showed in a constructive manner that $A^{G(\mathbb{C})}$ is a finitely generated algebra [13]. Our V corresponds with his 'binary forms of degree d', with $d = \dim V - 1$. Hilbert (1890) generalized Gordan's theorem nonconstructively to arbitrary n and – encouraged by an incorrect claim of Maurer – asked in his 14th problem to prove that this finite generation of invariants is a very general fact about actions of algebraic Lie groups on domains of finite type over $\mathbb{C}$. A counterexample of Nagata (1959) showed this was too optimistic, but by then it was understood that finite generation of invariants holds for compact connected real Lie groups (cf. Hurwitz 1897) as well as for their complexifications, also known as the connected reductive complex algebraic Lie groups (Weyl 1926). Finite groups have been treated by Emmy Noether (1926) [24], so connectedness may be dropped. (Algebraic Lie groups have finitely many connected components.)

Mumford (1965) needed finite generation of invariants for reductive algebraic groups over fields of arbitrary characteristic in order to construct moduli spaces. In his book Geometric Invariant Theory [22] he introduced a condition, often referred to as *geometric reductivity*, that he conjectured to be true for reductive algebraic groups and that he conjectured to imply finite generation of invariants. These conjectures were confirmed by Haboush (1975) [16] and Nagata (1964) [23] respectively. Nagata treated any algebra of finite type over the base field, not just domains. We adopt this generality. It rather changes the problem of finite generation of invariants. For instance, counterexamples to finite generation of invariants are now easy to find already when the Lie group $G(\mathbb{C})$ is $\mathbb{C}$ with addition as operation. (See Exercise 2.1.)

[We now understand that over an arbitrary commutative noetherian base ring the right counterpart of Mumford's geometric reductivity is not the geometric reductivity of Seshadri (1977) but the *power reductivity* of Franjou and van der Kallen (2010) [12], which is actually equivalent to the finite generation property.]

Let us say that G satisfies property (FG) if, whenever G acts on a commutative algebra of A finite type over k, the ring of invariants A^G is also finitely generated over k. So then the theorem of Haboush and Nagata says that connected reductive algebraic groups over a field have property (FG). Of course the action of G on A should be consistent with the nature of G and A respectively. Thus if G is an algebraic group, then the action should be algebraic and the multiplication map $A \otimes_k A \to A$ should be equivariant.

We will be interested in the cohomology algebra $H^*(G,A)$ of a geometrically reductive group G acting on a commutative algebra A of finite type over a base field k. Or, more generally, a power reductive affine flat algebraic group scheme G acting on a commutative algebra A of finite type over a noetherian commutative base ring k. Observe that $H^0(G,A)$ is just the algebra of invariants A^G, which we know to be finitely generated. The $H^i(G,-)$ are the right derived functors of the functor $V \mapsto V^G$.

My conjecture was that the full algebra $H^*(G,A)$ is finitely generated when k is field and G is a geometrically reductive group (or group scheme). Let us say that G satisfies the cohomological finite generation property (CFG) if, whenever G acts on a commutative algebra A of finite type over k, the cohomology algebra $H^*(G,A)$ is also finitely generated over k. So my conjecture was that if the base ring k is a field and an affine algebraic group (or group scheme) G over k satisfies property (FG) then it actually satisfies the stronger property (CFG). This was proved by Touzé [30], by constructing classes $c[m]$ in Ext groups in the category of *strict polynomial bifunctors* of Franjou and Friedlander [10]. If the base field has characteristic zero then there is little to do, because then (FG) implies that $H^{>0}(G,A)$ vanishes.

One may ask if (CFG) also holds when the base ring is not a field but just noetherian and $G = GL_n$ say. This question is still open for $n \geq 3$. But see [35].

We are not aware of striking applications of the general (CFG) theorem, but investigating the (CFG) conjecture has led to new insights [34]. The conjecture also fits into a long story where special cases have been very useful. The case of a finite group was treated by Evens (1961) [9] and this has been the starting point for the theory of *support varieties* [2, Chapter 5]. In this theory one exploits a connection between the rate of growth of a minimal projective resolution and the dimension of a 'support variety', which is a subvariety of the spectrum of $H^{\mathrm{even}}(G,k)$. The case of finite group schemes over a field (these are group schemes whose coordinate ring is a finite-dimensional vector space) turned out to be 'surprisingly elusive'. It was finally settled by Friedlander and Suslin (1997) [11]. For this they had to invent *strict polynomial functors* and compute with certain Ext groups in the category of strict polynomial functors. Again their result was crucial for developing a theory of support varieties, now for finite group schemes.

As $H^{>0}(G,k)$ vanishes for reductive G, there is no obvious theory of support varieties for reductive G.

Exercise 2.1 (Additive group is not reductive). Let $G = \mathbb{C}$ with addition as group operation. Make G act on $M = \mathbb{C}^2$ by $x \cdot (a,b) = (a + xb, b)$. Projection onto the second factor of $\mathbb{C}^2$ defines a surjective equivariant linear map $M \to \mathbb{C}$ with G acting trivially on the target. It induces a map of symmetric algebras $S^*_{\mathbb{C}}(M) \to S^*_{\mathbb{C}}(\mathbb{C})$. View $S^*_{\mathbb{C}}(\mathbb{C})$ as an $S^*_{\mathbb{C}}(M)$-module. Show that the algebra of invariants in the finite type $\mathbb{C}$-algebra $S^*_{S^*_{\mathbb{C}}(M)}(S^*_{\mathbb{C}}(\mathbb{C}))$ is not finitely generated. Hint: Exploit the trigrading.

Reductivity can be thought of as what one needs to avoid this example and its relatives. Reductivity of an affine algebraic group G over an algebraically closed field forbids that the connected component of the identity of G (for the Zariski topology) has a normal algebraic subgroup isomorphic to the additive group underlying a nonzero vector space. Originally reductivity referred to representations being completely reducible, but this meaning was abandoned in order to include groups over fields of positive characteristic that look pretty much like reductive groups over $\mathbb{C}$. For example GL_n is reductive, but when the ground field has positive characteristic, GL_n has representations that are not completely reducible. Indeed in positive characteristic the category of representations of GL_n has interesting Ext groups and this is our subject.

3. Some basic notions, notations and facts for group schemes

Let us now assume less familiarity with algebraic groups or group schemes.

3.1. Rings and algebras

Every ring has a unit and ring homomorphisms are unitary. Our *base ring* k is commutative noetherian and most of the time a field of characteristic $p > 0$, in fact just $\mathbb{F}_p$. Let Rg_k denote the category of commutative k-algebras. An object R of Rg_k is a commutative ring together with a homomorphism $k \to R$. We write $R \in \mathsf{Rg}_k$ to indicate that R is an object of Rg_k. The same convention will be used for other categories. When C is a category, C^{op} denotes the opposite category. Let Gp be the category of groups.

3.2. Group schemes

A functor $G : \mathsf{Rg}_k \to \mathsf{Gp}$ is called an affine flat algebraic *group scheme* over k if G is *representable* [36, 1.2], [19] by a flat k-algebra of finite type, which is then known as the *coordinate ring* $k[G]$ of G [7], [18], [36]. Recall that this means that for every R in Rg_k one is given a bijection between $\mathrm{Hom}_{\mathsf{Rg}_k}(k[G], R)$ and $G(R)$, thus providing $\mathrm{Hom}_{\mathsf{Rg}_k}(k[G], R)$ with a group structure, functorial in R. In particular one has the unit element $\epsilon : k[G] \to k$ in the group $G(k) \cong \mathrm{Hom}_{\mathsf{Rg}_k}(k[G], k)$. This ϵ is also known as the *augmentation map* of $k[G]$. In the group $\mathrm{Hom}_{\mathsf{Rg}_k}(k[G], k[G] \otimes_k k[G])$ one has the elements $x : f \mapsto f \otimes 1$ and $y : f \mapsto 1 \otimes f$ with product xy known as the *comultiplication* $\Delta_G : k[G] \to k[G] \otimes_k k[G]$. These maps ϵ, Δ_G make $k[G]$ into a *Hopf algebra* [36, 1.4]. (There is also an *antipode.*) If g, $h \in \mathrm{Hom}_{\mathsf{Rg}_k}(k[G], R)$ then gh in $G(R)$ is just $m_R \circ (g \otimes h) \circ \Delta_G$, where $m_R : R \otimes_k R \to R$ is the multiplication map of R.

3.3. G-modules

We will be working in the category Mod_G of G-modules. A G-module or *representation* of G is simply a *comodule* [36, 3.2] for the Hopf algebra $k[G]$. In functorial language this means that one is given a k-module V with an action of $G(R)$ on $V \otimes_k R$ by R-linear endomorphisms, functorially in the commutative k-algebra R.

In particular, the identity map $k[G] \to k[G]$ viewed as an element of $G(k[G])$ acts by a $k[G]$-linear map $V \otimes_k k[G] \to V \otimes_k k[G]$ and the composite of this $k[G]$-linear map with $v \mapsto v \otimes 1$ is the *comultiplication* $\Delta_V : V \to V \otimes_k k[G]$ defining the comodule structure of V. If $g \in \mathrm{Hom}_{\mathsf{Rg}_k}(k[G], R) \cong G(R)$, then it acts on $V \otimes_k R$ as $v \otimes r \mapsto (\mathrm{id} \otimes g)(\Delta_V(v))r$. The category Mod_G has useful properties only under the assumption that G is flat over k. That is why we always make this assumption. Flatness is of course automatic when k is a field. Geometers should be warned that it is a mistake to restrict attention to representations that are representable. So while our group functors are schemes, our representations need not be. For instance, in the (CFG) conjecture finite-dimensional algebras A are of less interest. And if A is infinite-dimensional as a vector space then as a representation it is no scheme.

3.4. Invariants

One may define the submodule V^G of fixed vectors or *invariants* of a representation V and get a natural isomorphism $\mathrm{Hom}_{\mathsf{Mod}_G}(k, V) \cong V^G$, where k also stands for the representation k^{triv} with underlying module k and trivial G action.

3.5. Cohomology of G-modules

The category Mod_G is abelian with enough injectives. We write Hom_G for $\mathrm{Hom}_{\mathsf{Mod}_G}$ and Ext_G for $\mathrm{Ext}_{\mathsf{Mod}_G}$. Cohomology is simply defined as follows:

$$H^i(G, V) := \mathrm{Ext}^i_G(k, V).$$

It may be computed [18, I 4.14–4.16] as the cohomology of the Hochschild complex $C^\bullet(V) = (V \otimes_k C^\bullet(k[G]))^G$. There is a *differential graded algebra* (=DGA) structure on $C^\bullet(k[G]) = k[G]^{\otimes(\bullet+1)}$. Let $R \in \mathsf{Rg}_k$ be provided with an action of G. So R is a G-module and the multiplication $R \otimes_k R \to R$ is a G-module map. If $u \in C^r(G, R)$ and $v \in C^s(G, R)$, then $u \cup v$ is defined in simplified notation by

$$(u \cup v)(g_1, \dots, g_{r+s}) = u(g_1, \dots, g_r).^{g_1 \cdots g_r} v(g_{r+1}, \dots, g_{r+s}),$$

where ${}^g r$ denotes the image of $r \in R$ under the action of g. With this cup product $C^*(G, R)$ is a differential graded algebra.

Remark 3.6. We have followed [18] in that we have used inhomogenous cochains, although for $C^\bullet(k[G])$ homogeneous cochains might be more natural. Thus one could take as alternative starting point a differential graded algebra $C^\bullet_{\mathrm{hom}}(k[G])$ with $C^i_{\mathrm{hom}}(k[G]) = k[G]^{\otimes(i+1)}$ and differential d as suggested by $(df)(g_0, g_1, g_2) = f(g_1, g_2) - f(g_0, g_2) + f(g_0, g_1)$. View $C^i_{\mathrm{hom}}(k[G])$ as G-module through left translation as in ${}^g f(g_0, \dots, g_i) = f(g^{-1}g_0, \dots, g^{-1}g_i)$. Then $H^i(G, V)$ may be computed as the cohomology of $(V \otimes_k C^\bullet_{\mathrm{hom}}(k[G]))^G$.

3.7. Symmetric and divided powers

For simplicity let k be a field. If V is a finite-dimensional vector space and $n \geq 1$, we have an action of the *symmetric group* $\mathfrak{S}_n$ on $V^{\otimes n}$ and the nth *symmetric power* $S^n(V)$ is the module of *coinvariants* [4, II 2] $(V^{\otimes n})_{\mathfrak{S}_n} = H_0(\mathfrak{S}_n, V^{\otimes n})$ for this action. Dually the nth *divided power* $\Gamma^n(V)$ is the module of invariants $(V^{\otimes n})^{\mathfrak{S}_n}$ [8], [27]. One has $\Gamma^n(V)^\vee \cong S^n(V^\vee)$.

Both S^* and Γ^* are *exponential functors*. That is, one has

$$S^n(V \oplus W) = \bigoplus_{i=0}^{n} S^i(V) \otimes_k S^{n-i}(W)$$

and similarly

$$\Gamma^n(V \oplus W) = \bigoplus_{i=0}^{n} \Gamma^i(V) \otimes_k \Gamma^{n-i}(W).$$

3.8. Tori

A very important example of an algebraic group scheme is the *multiplicative group* $\mathbb{G}_m$. It associates to R its group of invertible elements R^*. The coordinate ring $k[\mathbb{G}_m]$ is the Laurent polynomial ring $k[X, X^{-1}]$. Any $\mathbb{G}_m$-module V is a direct sum of *weight spaces* V_i on which Δ_V equals $v \mapsto v \otimes X^i$. Weight spaces are nonzero by definition.

Exercise 3.9. Prove this decomposition into weight spaces. Rewrite $k[X, X^{-1}] \otimes_k k[X, X^{-1}]$ as $k[X, X^{-1}, Y, Y^{-1}]$ where $X \otimes 1$ is written as X and $1 \otimes X$ as Y, so that $\Delta_{\mathbb{G}_m} X = XY$. Use that if $\Delta_V v = \sum_i \pi_i(v) X^i$, then $\sum_i \pi_i(v)(XY)^i = \sum_{i,j} \pi_j(\pi_i(v)) X^i Y^j$.

More generally the direct product T of r copies of $\mathbb{G}_m$, known as a *torus* T *of rank* r has as coordinate ring the Laurent polynomial ring in r variables $k[X_1, X_1^{-1}, \dots, X_r, X_r^{-1}]$. Again any T-module V is a direct sum of nonzero *weight spaces* V_λ where now the weight λ is an r-tuple of integers and Δ_V restricts to $v \mapsto v \otimes X_1^{\lambda_1} \cdots X_r^{\lambda_r}$ on V_λ. So a weight space is spanned by simultaneous eigenvectors with common eigenvalues and every T-module is diagonalizable. The invariants in a T-module are the elements of weight zero. Taking invariants is exact on Mod_T and $H^{>0}(T, V)$ always vanishes.

3.10. The additive group

The group scheme $\mathbb{G}_a$ sends a k-algebra R to the underlying additive group. The coordinate ring of $\mathbb{G}_a$ is $k[X]$ with $\Delta_{\mathbb{G}_a}(X) = X \otimes 1 + 1 \otimes X$. Recall that the additive group is not reductive. It has no property (FG). (Redo Exercise 2.1 with k replacing $\mathbb{C}$.) If k is a field of characteristic $p > 0$ then $H^1(\mathbb{G}_a, k)$ is already infinite-dimensional, so even with such small coefficient module the cohomology explodes. Thus cohomological finite generation is definitely tied with reductivity.

3.11. General linear group

Let $n \geq 1$. The group scheme GL_n associates to R the group $GL_n(R)$ of n by n matrices with entries in R and with invertible determinant. Its coordinate ring $k[GL_n]$ is $k[M_n][1/\det]$, where $k[M_n]$, also known as the coordinate ring of the *monoid* of n by n matrices, is the polynomial ring $k[X_{11}, X_{12}, \dots, X_{nn}]$ in n^2 variables $X_{11}, X_{12}, \dots, X_{nn}$ and det is the determinant of the matrix (X_{ij}). A ring homomorphism $\phi : k[M_n] \to R$ corresponds with the matrix $(\phi(X_{ij}))$ and ϕ extends to $k[GL_n]$ if and only if this matrix is invertible. One sees that indeed

$\mathrm{Hom}_{\mathsf{Rg}_k}(k[GL_n], R) \cong GL_n(R)$. If $n = 1$ we are back at $\mathbb{G}_m$, but as soon as $n \geq 2$ the representation theory becomes much more interesting. In fact there is a lemma (cf. [31, Lemma 1.7]) telling that for proving my (CFG) conjecture over a field k it suffices to show that the reductive group scheme $G = GL_n$ has (CFG), in particular for large n. The lemma explains why the homological algebra of strict polynomial bifunctors becomes so relevant: As we will see, it encodes what happens to $H^\bullet(GL_n, V_n)$ as n becomes large, for a certain kind of coefficients V_n.

3.12. Polynomial representations

Let k be a field until further notice. One calls a finite-dimensional representation V of GL_n a *polynomial representation* if the action is given by polynomials, meaning that Δ_V factors trough the embedding $V \otimes_k k[M_n] \to V \otimes_k k[GL_n]$. And one calls it homogeneous of *degree* d if moreover Δ_V lands in $V \otimes_k k[M_n]_d$, where $k[M_n]_d$ consists of polynomials homogeneous of total degree d. If one lets $\mathbb{G}_m$ act on $k[M_n]$ by algebra automorphisms giving the variables X_{ij} weight one and k weight zero, then $k[M_n]_d$ is just the weight space of weight d. Polynomial representations were studied by Schur in his thesis (1901). The *Schur algebra* $S_k(n, d)$ can be described as $\Gamma^d(\mathrm{End}_k(k^n))$ with multiplication obtained by restricting the usual algebra structure on $\mathrm{End}_k(k^n)^{\otimes d}$ given by $(f_1 \otimes \cdots \otimes f_d)(g_1 \otimes \cdots \otimes g_d) = f_1 g_1 \otimes \cdots \otimes f_d g_d$. The category of finitely generated left $S_k(n, d)$-modules is equivalent to the category of finite-dimensional polynomial representations of degree d of GL_n [11, §3].

3.13. Frobenius twist of a representation

Let p be a prime number and $k = \mathbb{F}_p$. The group scheme GL_n admits a Frobenius homomorphism $F : GL_n \to GL_n$ that sends a matrix $(a_{ij}) \in GL_n(R)$ to (a_{ij}^p). If V is a representation of GL_n then one gets a new representation $V^{(1)}$, called the *Frobenius twist*, by precomposing with F. If V is a polynomial representation of degree d then $V^{(1)}$ has degree pd. One may also twist r times and obtain $V^{(r)}$. We do not reserve the notation F for Frobenius, but $V^{(r)}$ will always indicate an r-fold Frobenius twist.

Exercise 3.14. We keep $k = \mathbb{F}_p$. Let V be a finite-dimensional representation of GL_n. Choose a basis in V. The action of $g \in GL_n(R)$ on $V \otimes_k R$ is given with respect to the chosen basis by a matrix (g_{ij}) with entries in R. Show that the action on $V^{(r)} \otimes_k R$ is given by the matrix $(g_{ij}^{p^r})$. In other words, when the base field is $\mathbb{F}_p$ one may confuse precomposition by Frobenius with postcomposition. For larger ground fields one would have to be more careful.

4. Some basic notions, notations and facts for functors

4.1. Strict polynomial functors

Let $\mathcal{V}_k$ be the k-linear category of finite-dimensional vector spaces over a field k. The category $\Gamma^d\mathcal{V}_k$, often written $\Gamma^d\mathcal{V}$, generalizes the Schur algebras as follows. Its objects are finite-dimensional vector spaces over k, but $\mathrm{Hom}_{\Gamma^d\mathcal{V}}(V, W) =$

$\Gamma^d(\mathrm{Hom}_k(V,W))$. The composition is similar to the one in a Schur algebra. We could call $\Gamma^d\mathcal{V}_k$ the *Schur category.* The category of *strict polynomial functors of degree* d is now defined, following the exposition of Pirashvili [25], [26], as the category of k-linear functors $\Gamma^d\mathcal{V} \to \mathcal{V}_k$. The reason for the word *strict* is simply that the terminology *polynomial functor* already means something. There is an obvious functor ι^d from $\mathcal{V}_k$ to $\Gamma^d\mathcal{V}$. It sends $V \in \mathcal{V}_k$ to $V \in \Gamma^d\mathcal{V}$ and $f \in \mathrm{Hom}_k(V,W)$ to $f^{\otimes d}$. This is not k-linear when $d > 1$. If $F \in \mathcal{P}_d$, let us try to understand the composite map $\mathrm{Hom}_k(V,W) \to \mathrm{Hom}_k(FV,FW)$. The map $\mathrm{Hom}_{\Gamma^d\mathcal{V}}(V,W) \to \mathrm{Hom}_k(FV,FW)$ is k-linear and is thus given by an element ψ of the space $\mathrm{Hom}_k(\Gamma^d(\mathrm{Hom}_k(V,W)),\mathrm{Hom}_k(FV,FW)) \cong \mathrm{Hom}_k(FV,FW) \otimes S^d(\mathrm{Hom}_k(V,W)^\vee)$ which also encodes the polynomial maps from $\mathrm{Hom}_k(V,W)$ to $\mathrm{Hom}_k(FV,FW)$ that are homogeneous of degree d. One checks that the composite map $\mathrm{Hom}_k(V,W) \to \mathrm{Hom}_k(FV,FW)$ is the polynomial map of degree d encoded by ψ. This explains why F is called a (strict) polynomial functor of degree d.

Remark 4.2. The original definition of Friedlander and Suslin did not use $\Gamma^d\mathcal{V}$, but just defined strict polynomial functors of degree d as functors $F : \mathcal{V}_k \to \mathcal{V}_k$ enriched with elements $\phi_{V,W}$ in $\mathrm{Hom}_k(FV,FW) \otimes S^d(\mathrm{Hom}_k(V,W)^\vee)$ that satisfy appropriate conditions, like the condition that the polynomial map $\mathrm{Hom}_k(V,W) \to \mathrm{Hom}_k(FV,FW)$ encoded by $\phi_{V,W}$ agrees with F. That is more intuitive, but the definition by means of $\Gamma^d\mathcal{V}$ is concise and has its own advantages. In fact one may view $F : \Gamma^d\mathcal{V} \to \mathcal{V}_k$ as exactly the enrichment that Friedlander and Suslin need to add to the composite functor $F\iota^d$. One should use both points of view. They are equivalent [25]. We will secretly think in terms of the Friedlander and Suslin setting when that is more convenient.

4.3. Some examples of strict polynomial functors

The functor $F = \otimes^d$ maps $V \in \Gamma^d\mathcal{V}$ to $V^{\otimes d}$. If $f \in \mathrm{Hom}_{\Gamma^d\mathcal{V}}(V,W)$, view f as an element of $\mathrm{Hom}_k(V,W)^{\otimes d}$ and define $Ff : FV \to FW$ by means of the pairing $\mathrm{Hom}_k(V,W)^{\otimes d} \times V^{\otimes d} \to W^{\otimes d}$ which maps the pair $(f_1 \otimes \cdots \otimes f_d, v_1 \otimes \cdots \otimes v_d)$ to $f_1(v_1) \otimes \cdots \otimes f_d(v_d)$.

The functor Γ^d is the subfunctor of $\otimes^d$ with value $\Gamma^d(V)$ on $V \in \Gamma^d\mathcal{V}$.

The functor S^d is the quotient functor of $\otimes^d$ with value $S^d(V)$ on $V \in \Gamma^d\mathcal{V}$.

If $F \in \mathcal{P}_d$, then its *Kuhn dual* $F^\#$ is defined as DFD, where $DV = V^\vee$ is the contravariant functor on $\mathcal{V}_k$ or $\Gamma^d\mathcal{V}_k$ sending V to its k-linear dual $V^\vee$. Thus $S^{d\#} = \Gamma^d$.

If k has characteristic $p > 0$, then the rth *Frobenius twist functor* $I^{(r)} \in \mathcal{P}_{p^r}$ is the subfunctor of S^{p^r} such that the vector space $I^{(r)}V$ is generated by the $v^{p^r} \in S^{p^r}V$. Note that every element of $I^{(r)}V$ is actually of the form v^{p^r} if $k = \mathbb{F}_p$.

4.4. Polynomial representations from functors

If $F \in \mathcal{P}_d$ then $F(k^n)$ is a polynomial representation of degree d of GL_n. The comodule structure is obtained from the homomorphism

$$\mathrm{Hom}_{\Gamma^d\mathcal{V}}(k^n,k^n) \to \mathrm{Hom}(F(k^n),F(k^n))$$

by means of the isomorphism

$$\begin{aligned}&\mathrm{Hom}_k(\mathrm{Hom}_{\Gamma^d\mathcal{V}}(k^n,k^n),\mathrm{Hom}(F(k^n),F(k^n)))\\&\quad\cong \mathrm{Hom}(F(k^n),F(k^n))\otimes_k S^d(\mathrm{Hom}_k(k^n,k^n)^\vee).\end{aligned}$$

Friedlander and Suslin showed [11, §3] that if $n \geq d$ this actually provides an equivalence of categories, preserving Ext groups [11, Cor 3.12.1], between $\mathcal{P}_d$ and the category of finite-dimensional polynomial representations of degree d of GL_n. So again there is another way to look at $\mathcal{P}_d$ and we secretly think in terms of polynomial representations when we find that more convenient.

Exercise 4.5 (Polarization). If k is a finite field and $V \in \mathcal{V}_k$, then V is a finite set. If $\dim V > 1$ and d is large, then the dimension of $\Gamma^d V$ exceeds the number of elements of V so that $\Gamma^d V$ is certainly not spanned by elements of the form $v^{\otimes d}$. On the other hand Friedlander and Suslin show that $\Gamma^d V$ is spanned by such elements if k is big enough, when keeping d and $\dim V$ fixed. So as long as one uses constructions that are compatible with base change one may think of $\Gamma^d V$ as spanned by the $v^{\otimes d}$.

Let $V = k^n$. Show that the $v^{\otimes d}$ generate $\Gamma^d V$ as a GL_n-module. Hint: Let T be the group scheme of diagonal matrices in GL_n. Show that the weight spaces of T in $\Gamma^d V$ are one-dimensional. Any GL_n-submodule must be a T-submodule, hence a sum of weight spaces. Now compute the weight decomposition of $v^{\otimes d}$ for $v \in V$.

4.6. Composition of strict polynomial functors

If $F \in \mathcal{P}_d$, $G \in \mathcal{P}_e$ we wish to define their composite $F \circ G \in \mathcal{P}_{de}$. Associated to F one has the functor $F\iota^d : \mathcal{V}_k \to \mathcal{V}_k$ and associated to G one has $G\iota^e : \mathcal{V}_k \to \mathcal{V}_k$. We want $F \circ G$ to correspond with the composite of $F\iota^d$ and $G\iota^e$. For $V \in \Gamma^{de}\mathcal{V}$ one puts $(F \circ G)V = F(GV)$. For $f \in \mathrm{Hom}_k(V, W)$ we want that $(F \circ G)f^{\otimes de}$ equals $F(G(f^{\otimes e})^{\otimes d})$. Thus let $\phi : \mathrm{Hom}_{\Gamma^e\mathcal{V}}(V, W) \to \mathrm{Hom}_k(GV, GW)$ be given by G and observe that the restriction of $\Gamma^d\phi : \Gamma^d \mathrm{Hom}_{\Gamma^e\mathcal{V}}(V, W) \to \Gamma^d \mathrm{Hom}_k(GV, GW)$ to $\Gamma^{de} \mathrm{Hom}_k(V, W)$ lands in the source of the map

$$\mathrm{Hom}_{\Gamma^d\mathcal{V}}(GV, GW) \to \mathrm{Hom}_k(FGV, FGW).$$

Exercise 4.7. Finish the definition and check all details.

In particular, the composite $F \circ I^{(r)}$ is called the rth *Frobenius twist* $F^{(r)}$ *of the functor* $F \in \mathcal{P}_d$. Recall that if $n \geq d$ we have an equivalence of categories, between $\mathcal{P}_d$ and the category of finite-dimensional polynomial representations of GL_n of degree d. Take $k = \mathbb{F}_p$ for simplicity. Now check that the notion of Frobenius twist on the strict polynomial side agrees with the notion of Frobenius twist for representations.

4.8. Untwist

For F, $G \in \mathcal{P}_d$ and $r \geq 1$ we have $\mathrm{Hom}_{\mathcal{P}_d}(F, G) \cong \mathrm{Hom}_{\mathcal{P}_{dp^r}}(F^{(r)}, G^{(r)})$ by [32, Lemma 2.2]. So to construct a morphism in $\mathcal{P}_d$ one may twist first. This was well known in the context of representations of GL_n, but the proof we know there involves fppf sheaves [18, I 9.5; I 6.3].

On Ext_{GL_n} groups Frobenius twist gives injective maps [18, II 10.14], but often no isomorphisms. Compare the formality conjecture below. In view of the connection between Ext_{GL_n} groups and $\mathrm{Ext}_{\mathcal{P}_d}$ groups we may also state this twist injectivity as

Theorem 4.9 (Twist Injectivity). *Let $F, G \in \mathcal{P}_d$. Precomposition by $I^{(1)}$ induces an injective map $\mathrm{Ext}^i_{\mathcal{P}_d}(F,G) \to \mathrm{Ext}^i_{\mathcal{P}_{dp}}(F^{(1)}, G^{(1)})$ for every $i \geq 0$.*

4.10. Parametrized functors

If $V \in \Gamma^d\mathcal{V}$ define the functor $V \otimes_k^{\Gamma^d} - : \Gamma^d\mathcal{V} \to \Gamma^d\mathcal{V}$ by sending an object W to $V \otimes_k W$ and a morphism $f \in \mathrm{Hom}_{\Gamma^d\mathcal{V}}(W,Z)$ to its image under $\Gamma^d(\phi)$: $\Gamma^d \mathrm{Hom}_k(W,Z) \to \Gamma^d \mathrm{Hom}_k(V \otimes_k W, V \otimes_k Z)$ where $\phi : g \mapsto \mathrm{id}_V \otimes g$. One checks that $V \otimes_k^{\Gamma^d} -$ is functorial in V.

If $F \in \mathcal{P}_d$, $V \in \Gamma^d\mathcal{V}$, then F_V denotes the composite $F(V \otimes_k^{\Gamma^d} -)$, $F_V W = F(V \otimes_k^{\Gamma^d} W)$. It is covariantly functorial in V, which is why we use a subscript. Dually, F^V denotes $F(\mathrm{Hom}_k(V,-)) = ((F^\#)_V)^\#$. It is contravariantly functorial in V, which is why we use a superscript. Notice that we did not decorate Hom_k with Γ^d like we did with $\otimes_k$. We leave that to the reader.

For example, $\Gamma^{dV} W = \Gamma^d(\mathrm{Hom}_k(V,W)) = \mathrm{Hom}_{\Gamma^d\mathcal{V}}(V,W)$, so that the Yoneda lemma [36, 1.3], [19] gives

$$\mathrm{Hom}_{\mathcal{P}_d}(\Gamma^{dV}, F) \cong FV.$$

As FV is exact in F, it follows that Γ^{dV} is projective in $\mathcal{P}_d$. Dually $S^d_V = \Gamma^{dV\#}$ is injective in $\mathcal{P}_d$ and

$$\mathrm{Hom}_{\mathcal{P}_d}(F, S^d_V) \cong F^\#(V).$$

4.11. An adjunction

For F, $G \in \mathcal{P}_d$ we have

$$\mathrm{Hom}_{\mathcal{P}_d}(F^V, G) \cong \mathrm{Hom}_{\mathcal{P}_d}(F, G_V)$$

in $\mathcal{P}_d$. So $F \mapsto F^V$ has *right adjoint* [19] $G \mapsto G_V$. Indeed the standard map $V \otimes_k \mathrm{Hom}_k(V,W) \to W$ in $\mathcal{V}_k$ induces a morphism $V \otimes_k^{\Gamma^d} \mathrm{Hom}_k(V,W) \to W$ in $\Gamma^d\mathcal{V}$ so that if $\phi : F \to G_V$ one gets a map $F^V W = F\,\mathrm{Hom}_k(V,W) \to G_V \mathrm{Hom}_k(V,W) \to GW$, functorial in G. If $G = S^d_Z$ then $\mathrm{Hom}_{\mathcal{P}_d}(F, G_V) \to \mathrm{Hom}_{\mathcal{P}_d}(F^V, G)$ becomes the isomorphism $F^\#(Z \otimes_k^{\Gamma^d} V) \to F^\#(V \otimes_k^{\Gamma^d} Z)$. As $\mathrm{Hom}_{\mathcal{P}_d}(-,-)$ is left exact, the result follows from this and functoriality in G.

4.12. Coresolutions

If $\dim V \geq d$ then Γ^{dV} forms a *projective generator* [19] of $\mathcal{P}_d$ and S^d_V an injective cogenerator. Say $V = k^n$ with $n \geq d$ and let $G = GL_n$ again. One may also write $G = GL_V$. For $F \in \mathcal{P}_d$ we have $FW \cong \mathrm{Hom}_{\mathcal{P}_d}(F^\#, S^d_W) \cong \mathrm{Hom}_G(F^\# V, S^d_W V) \hookrightarrow \mathrm{Hom}_k(F^\# V, S^d_W V) \cong \mathrm{Hom}_k(F^\# V, S^d_V W)$, functorially in W, so that

$$F \hookrightarrow \mathrm{Hom}_k(F^\# V, S^d_V).$$

And $\mathrm{Hom}_k(F^{\#}V, S^d_V)$ is just a direct sum of $\dim F^{\#}V$ copies of S^d_V, so it is injective and we conclude that $\mathcal{P}_d$ has enough injectives. Therefore we know now how to build injective coresolutions consisting of direct sums of copies of S^d_V. As $\mathrm{End}_{\mathcal{P}_d}(S^d_V) \cong \mathrm{End}_{\Gamma^d V}(V)$ we also have a grip on the differentials in these coresolutions.

So far we discussed coresolving an object of $\mathcal{P}_d$. We also want to coresolve cochain complexes. When we speak of a *cochain complex* $C^\bullet$ we do not assume C^i to vanish for $i < 0$. When f is a cochain map, we may use the symbol $\hookrightarrow$ to indicate that is an injective cochain map. If

$$C^\bullet = \cdots \stackrel{d}{\to} C^{-1} \stackrel{d}{\to} C^0 \stackrel{d}{\to} C^1 \stackrel{d}{\to} \cdots$$

is a cochain complex in $\mathcal{P}_d$ then one may find an injective cochain map $C^\bullet \hookrightarrow J^\bullet$ with each J^i injective and J^i zero when C^i is zero. This is clear when $C^\bullet$ is an easy complex like $\cdots \to 0 \to F \to 0 \to \cdots$ or $\cdots \to 0 \to F \stackrel{\mathrm{id}}{\to} F \to 0 \cdots$. Any $C^\bullet$ can be embedded into a direct sum of such easy complexes.

Recall that a cochain map $f : C^\bullet \to D^\bullet$ is called a *quasi-isomorphism* if each $H^i(f) : H^i(C^\bullet) \to H^i(D^\bullet)$ is an isomorphism.

If $C^\bullet$ is a cochain complex in $\mathcal{P}_d$ that is *bounded below*, meaning that $C^j = 0$ for $j \ll 0$, then one may find a quasi-isomorphism $C^\bullet \hookrightarrow J^\bullet$ with each J^j injective and J^j zero when $j \ll 0$. One may construct $J^\bullet$ as the total complex of a double complex $K^\bullet_\bullet$ obtained by coresolving like this: Construct an exact complex of complexes $0 \to C^\bullet \to K^\bullet_0 \to K^\bullet_1 \to \cdots$ where the $K^\bullet_i$ are complexes of injectives with $K^j_i = 0$ when $C^j = 0$. (Our double complexes commute so that a total complex requires appropriate signs.) One calls $C^\bullet \hookrightarrow J^\bullet$, or simply $J^\bullet$, an *injective coresolution* of $C^\bullet$. Notice that we prefer our injective coresolutions to be injective as cochain maps, as indicated by the symbol $\hookrightarrow$. But *any* quasi-isomorphism $C^\bullet \to J^\bullet$ with each J^j injective is called an injective coresolution of $C^\bullet$.

If $f : J^\bullet \to \tilde{J}^\bullet$ is a quasi-isomorphism of bounded below complexes of injectives, then the mapping cone of f is a bounded below acyclic complex of injectives, hence split and contractible, and f is a homotopy equivalence [4, Proposition 0.3, Proposition 0.7].

Remark 4.13. Actually $\mathcal{P}_d$ has *finite global dimension* [37] by [18, A.11] so that even for an unbounded complex $C^\bullet$ there is a quasi-isomorphism $C^\bullet \hookrightarrow J^\bullet$ with each J^j injective. Indeed the coresolution $0 \to C^\bullet \to K^\bullet_0 \to K^\bullet_1 \to \cdots$ may be terminated and thus one may use the total complex of the finite width double complex $K^\bullet_0 \to K^\bullet_1 \to \cdots \to K^\bullet_M$. Also, a bounded complex is quasi-isomorphic to a bounded complex of injectives. (A complex $C^\bullet$ is called *bounded* if C^i vanishes for $|i| \gg 0$.) Passing to Kuhn duals one also sees that a bounded complex is quasi-isomorphic to a complex of projectives that is bounded.

Exercise 4.14. Let $C^\bullet \hookrightarrow D^\bullet$ be a quasi-isomorphism and let $C^\bullet \to E^\bullet$ be a cochain map. Then $E^\bullet \hookrightarrow (D^\bullet \oplus E^\bullet)/C^\bullet$ is a quasi-isomorphism. There is a commutative diagram

$$\begin{array}{ccc} C^\bullet & \hookrightarrow & D^\bullet \\ \downarrow & & \downarrow \\ E^\bullet & \hookrightarrow & F^\bullet \end{array}$$

with $E^\bullet \hookrightarrow F^\bullet$ a quasi-isomorphism. (Hint: One may also construct an anticommutative square.) We will refer to this diagram as the *base change* diagram.

Let $\mathcal{K}^+\mathcal{P}_d$ be the homotopy category [37, Exercise 1.4.5] of bounded below cochain complexes in $\mathcal{P}_d$. If $f : J^\bullet \to C^\bullet$ is a quasi-isomorphism of bounded below complexes and the J^i are injectives, then f defines a split monomorphism in $\mathcal{K}^+\mathcal{P}_d$.

The injective coresolution of a bounded below complex is unique up to homotopy equivalence. Here one does not require the coresolutions to be injective as cochain maps. (But to prove it, consider a pair of injective coresolutions with at least one of the two injective as cochain map. Then use base change and coresolve.)

Let $A^\bullet$ be an exact bounded below complex. Its injective coresolutions are contractible. If $f : A^\bullet \to \tilde{J}^\bullet$ is a cochain map and $\tilde{J}^\bullet$ is a bounded below complexes of injectives, then f is homotopic to zero. (Hint: Take $C^\bullet = A^\bullet$ and $E^\bullet = \tilde{J}^\bullet$ in the base change diagram.)

Let $f : J^\bullet \to \tilde{J}^\bullet$ be a morphism of bounded below complexes of injectives. If there is a quasi-isomorphism $g : C^\bullet \hookrightarrow J^\bullet$ so that $fg = 0$, then f factors through $J^\bullet/C^\bullet$ and is thus homotopic to zero.

Definition 4.15. We say that two complexes $C^\bullet$, $D^\bullet$ are *quasi-isomorphic* if there are complexes $E_0^\bullet, \dots, E_{2n}^\bullet$ and quasi-isomorphisms $f_i : E_{2i}^\bullet \to E_{2i+1}^\bullet$, $g_i : E_{2i+2}^\bullet \to E_{2i+1}^\bullet$ with $E_0^\bullet = C^\bullet$, $E_{2n}^\bullet = D^\bullet$. Thus $C^\bullet$, $D^\bullet$ are joined by zigzags of quasi-isomorphisms $E_{2i}^\bullet \to E_{2i+1}^\bullet \leftarrow E_{2i+2}^\bullet$.

It follows from Exercise 4.14 that injective coresolutions of quasi-isomorphic bounded below complexes are homotopy equivalent. This fact will underlie our choice of model for the derived category (*cf.* [37, Theorem 10.4.8]).

5. Precomposition by Frobenius

We will need the derived category $\mathcal{D}^b\mathcal{P}_d$ to discuss the formality conjecture of Chałupnik, which is formulated in terms of $\mathcal{D}^b\mathcal{P}_d$. In 6.8 we will turn to the collapsing conjecture of Touzé. Its formulation and proof do not need anything about derived categories. That part of the story can be told entirely on the level of spectral sequences of bicomplexes, but we leave it to the reader to disentangle the derived categories from the spectral sequences. We find the analogy between the formality problem and the collapsing conjecture instructive. Our use of derived categories is rather basic. We will model the derived category $\mathcal{D}^b\mathcal{P}_d$ by a certain homotopy category $\mathcal{K}^b\mathcal{I}_d$. We could have phrased almost everything in terms of that homotopy category, but we like derived categories and their intimate connections with spectral sequences.

On closer inspection the reader will find that even the existence of $\mathcal{D}^b\mathcal{P}_d$ is not essential for the heart of the arguments. One may simply view $\mathcal{D}^+\mathcal{P}_d$ as a source of inspiration and notation.

5.1. Derived categories

One gets the *derived category* $\mathcal{D}^+\mathcal{P}_d$ from the category of bounded below cochain complexes in $\mathcal{P}_d$ by forcing quasi-isomorphic complexes to be isomorphic. There are several ways to do that. The usual way is by formally inverting the quasi-isomorphisms. (The objects of the category do not change. Only the morphism sets are changed when throwing in formal inverses.)

In our case there is a good alternative: Replace every complex by an injective coresolution, then compute up to homotopy. In fact, if $C^\bullet$, $D^\bullet$ are bounded below cochain complexes and $D^\bullet \hookrightarrow \tilde{J}^\bullet$ is an injective coresolution, then $\mathrm{Hom}_{\mathcal{D}^+\mathcal{P}_d}(C^\bullet, D^\bullet)$ may be identified by [37, Cor 10.4.7] with $\mathrm{Hom}_{\mathcal{K}^+\mathcal{P}_d}(C^\bullet, \tilde{J}^\bullet)$, where $\mathcal{K}^+\mathcal{P}_d$ is the homotopy category [37, Exercise 1.4.5] of bounded below cochain complexes in $\mathcal{P}_d$. One does not need to coresolve $C^\bullet$ here, but one may coresolve it too. If $J^\bullet$ is an injective coresolution of $C^\bullet$, then $\mathrm{Hom}_{\mathcal{K}^+\mathcal{P}_d}(C^\bullet, \tilde{J}^\bullet)$ is isomorphic to $\mathrm{Hom}_{\mathcal{K}^+\mathcal{P}_d}(J^\bullet, \tilde{J}^\bullet)$.

Note that a cochain map $f : C^\bullet \to D^\bullet$ is homotopic to zero if and only if it factors through the mapping cone of $\mathrm{id} : D^\bullet \to D^\bullet$. And this mapping cone is quasi-isomorphic to the zero complex. Taking into account the k-linear structure it is thus not surprising that inverting quasi-isomorphisms forces homotopic cochain maps to become equal [37, Examples 10.3.2]. The derived category may also be described [37, 10.3] by first passing to $\mathcal{K}^+\mathcal{P}_d$ and then inverting the quasi-isomorphisms.

Consider the full subcategory $\mathcal{K}^+\mathcal{I}_d$ of the homotopy category whose objects are bounded below complexes of injectives in $\mathcal{P}_d$. It maps into $\mathcal{D}^+\mathcal{P}_d$ and $\mathcal{K}^+\mathcal{I}_d \to \mathcal{D}^+\mathcal{P}_d$ is an equivalence of categories [37, Theorem 10.4.8]. One retracts $\mathcal{D}^+\mathcal{P}_d$ back to $\mathcal{K}^+\mathcal{I}_d$ by sending a complex $C^\bullet$ to an injective coresolution $J^\bullet$ of $C^\bullet$. We use $\mathcal{K}^+\mathcal{I}_d$ as our working definition of $\mathcal{D}^+\mathcal{P}_d$. Note that the definition of $\mathcal{K}^+\mathcal{I}_d$ is easy. No formal inverting is needed. The way it works is that, when we try to understand morphisms in $\mathcal{D}^+\mathcal{P}_d$, we may model an object of $\mathcal{D}^+\mathcal{P}_d$ by means of its image under the retract.

We view $\mathcal{P}_d$ as a subcategory of $\mathcal{D}^+\mathcal{P}_d$ in the usual way: Associate to $F \in \mathcal{P}_d$ [an injective coresolution of] the complex $\cdots \to 0 \to F \to 0 \to \cdots$ with F in degree zero. Write the complex as $F[-m]$ when F is placed in degree m instead. The derived category encodes Ext groups as follows [37, 10.7]. If F, $G \in \mathcal{P}_d$ then

$$\mathrm{Hom}_{\mathcal{D}^+\mathcal{P}_d}(F[-m], G[-n]) \cong \mathrm{Ext}^{m-n}_{\mathcal{P}_d}(F, G).$$

One also has the *bounded derived category* $\mathcal{D}^b\mathcal{P}_d$ which we think of as the full subcategory of $\mathcal{D}^+\mathcal{P}_d$ whose objects C have vanishing $H^i(C)$ for $i \gg 0$. Let $\mathcal{K}^b\mathcal{I}_d$ be the subcategory of $\mathcal{K}^+\mathcal{I}_d$ whose objects are homotopy equivalent to a bounded complex of injectives in $\mathcal{P}_d$. Then $\mathcal{K}^b\mathcal{I}_d$ is our working definition of $\mathcal{D}^b\mathcal{P}_d$. (Recall that $\mathcal{P}_d$ has finite cohomological dimension.)

5.2. The adjoint of the twist

We now aim for a formality conjecture of Chałupnik [6] related to the collapsing conjecture of Touzé [32, Conjecture 8.1]. These conjectures imply a powerful formula (Exercise 5.8) for the effect of Frobenius twist on Ext groups in the category of strict polynomial functors. For the application to the (CFG) conjecture we will need to extend the theory from strict polynomial functors to strict polynomial bifunctors, but the difficulties are already visible for strict polynomial functors.

Let $A \in \mathcal{P}_e$. The example we have in mind is $A = I^{(r)}$, the rth Frobenius twist. Precomposition with A defines a functor $\mathcal{P}_d \to \mathcal{P}_{de} : F \mapsto F \circ A$. So the example we have in mind is $F \mapsto F^{(r)}$. The functor $F \mapsto F \circ A$ extends to a functor $- \circ A : \mathcal{K}^b\mathcal{I}_d \to \mathcal{D}^b\mathcal{P}_{de}$, hence a functor $\mathcal{D}^b\mathcal{P}_d \to \mathcal{D}^b\mathcal{P}_{de}$. We first seek its right adjoint $\mathbf{K}^{\mathbf{r}}_A$. For an object $J^\bullet$ of $\mathcal{K}^b\mathcal{I}_{de}$ put

$$\mathbf{K}^{\mathbf{r}}_A(J^\bullet)(V) := \mathrm{Hom}_{\mathcal{P}_{de}}(\Gamma^{dV} \circ A, J^\bullet),$$

where the right-hand side is viewed as a complex in $\mathcal{P}_d$ of functors

$$V \mapsto \mathrm{Hom}_{\mathcal{P}_{de}}(\Gamma^{dV} \circ A, J^i).$$

Observe that this complex is homotopy equivalent to a bounded complex. If $G \in \mathcal{D}^b\mathcal{P}_{de}$, then we take an injective coresolution $J^\bullet$ of G and put $\mathbf{K}^{\mathbf{r}}_A(G) := \mathbf{K}^{\mathbf{r}}_A(J^\bullet)$. Our claim is that

$$\mathrm{Hom}_{\mathcal{D}^b\mathcal{P}_{de}}(F \circ A, G) = \mathrm{Hom}_{\mathcal{D}^b\mathcal{P}_d}(F, \mathbf{K}^{\mathbf{r}}_A(G)),$$

for $F \in \mathcal{D}^b\mathcal{P}_d$, $G \in \mathcal{D}^b\mathcal{P}_{de}$.

Now take $Z \in \Gamma^{de}\mathcal{V}$ of dimension at least de. Then every object in $\mathcal{D}^+\mathcal{P}_{de}$ is quasi-isomorphic to one of the form

$$G = \cdots \to k^{n_i} \otimes_k S^{de}_Z \to k^{n_{i+1}} \otimes_k S^{de}_Z \to \cdots,$$

so we may assume that G is actually of this form. Notice that $k^{n_i} \otimes_k S^{de}_Z \to k^{n_{i+1}} \otimes_k S^{de}_Z$ is given by an n_{i+1} by n_i matrix with entries in $\mathrm{End}_{\Gamma^{de}\mathcal{V}}(Z)$. We may also assume $F = F^\bullet$ consists of projectives and is bounded. (We wish to use *balancing* [37, 2.7], which is the principle that both projective resolutions and injective coresolutions may be used to compute 'hyper Ext'. See also Section 5.13. We do not use an injective coresolution of F.) Put $F_i = F^{-i}$. Now $\mathrm{Hom}_{\mathcal{D}^+\mathcal{P}_{de}}(F \circ A, G)$ is computed as the H^0 of the total complex associated to the bicomplex $\mathrm{Hom}_{\mathcal{P}_{de}}(F_i \circ A, G^j)$ and $\mathrm{Hom}_{\mathcal{D}^b\mathcal{P}_d}(F, \mathbf{K}^{\mathbf{r}}_A(G))$ is similarly computed by way of a bicomplex [37, 2.7.5, Cor 10.4.7]. So let us compare the bicomplexes. We have $\mathrm{Hom}_{\mathcal{P}_{de}}(F_i \circ A, G^j) = \mathrm{Hom}_{\mathcal{P}_{de}}(F_i \circ A, k^{n_j} \otimes_k S^{de}_Z) = k^{n_j} \otimes_k \mathrm{Hom}_{\mathcal{P}_{de}}(F_i \circ A, S^{de}_Z) = k^{n_j} \otimes_k (F_i \circ A)^\# Z$ and $\mathrm{Hom}_{\mathcal{P}_d}(F_i, \mathbf{K}^{\mathbf{r}}_A(G)^j) = \mathrm{Hom}_{\mathcal{P}_d}(F_i, V \mapsto \mathrm{Hom}_{\mathcal{P}_{de}}(\Gamma^{dV} \circ A, k^{n_j} \otimes_k S^{de}_Z)) = k^{n_j} \otimes_k \mathrm{Hom}_{\mathcal{P}_d}(F_i, (V \mapsto (\Gamma^{dV} \circ A)^\# Z)) = k^{n_j} \otimes_k \mathrm{Hom}_{\mathcal{P}_d}(F_i, S^d_{A^\# Z}) = k^{n_j} \otimes_k F_i^\# A^\# Z$. The claim follows. (Exercise.)

Remark 5.3. These bicomplexes $\mathrm{Hom}_{\mathcal{P}_{de}}(F_i \circ A, G^j)$ and $\mathrm{Hom}_{\mathcal{P}_d}(F_i, \mathbf{K}^{\mathbf{r}}_A(G)^j)$ are meaningful by themselves. The fact that their total complexes are quasi-isomorphic may also serve as motivation for the definition of $\mathbf{K}^{\mathbf{r}}_A(G)$. This does not explicitly involve the derived category. It is closer in spirit to Section 5.13. The bicomplexes do not require that $F^\bullet$ is a bounded complex of projectives. A bounded above complex of projectives would do.

5.4. Formality

A bounded below cochain complex $C^\bullet$ in $\mathcal{P}_d$ is called *formal* if it is isomorphic in the derived category $\mathcal{D}^+\mathcal{P}_d$ to a complex $E^\bullet$ with zero differential. Notice that here we do not replace $E^\bullet$ with an injective coresolution, because that usually spoils the vanishing of the differential. Notice also that $E^i \cong H^i(E^\bullet) \cong H^i(C^\bullet)$. One can show that the isomorphism in the derived category is given by a single zigzag of quasi-isomorphisms $C^\bullet \to D^\bullet \leftarrow E^\bullet$ or, dually, a single zigzag of quasi-isomorphisms $C^\bullet \leftarrow D^\bullet \to E^\bullet$. One may also define a cochain complex to be formal if it is quasi-isomorphic in the sense of Definition 4.15 to a complex with zero differential. So one does not need the derived category to introduce formality.

Exercise 5.5. A complex has differential zero if and only if it is a direct sum of complexes each of which is concentrated in one degree. Let m be an integer. Let $C^\bullet$ be a bounded below cochain complex with $H^i(C^\bullet) = 0$ for $i \neq m$. Show that $C^\bullet$ is formal by constructing a zigzag of quasi-isomorphisms $C^\bullet \hookleftarrow D^\bullet \twoheadrightarrow E^\bullet$ where $D^i = 0$ for $i > m$, $E^i = 0$ for $i \neq m$.

Remark 5.6. Let the 2-fold extension $0 \to F \to G \xrightarrow{f} H \to K \to 0$ represent [1, 2.6] a nonzero element of $\mathrm{Ext}^2_{\mathcal{P}_d}(K, F)$. One can show that

$$\cdots \to 0 \to G \xrightarrow{f} H \to 0 \to \cdots$$

is not formal.

For example, consider the 2-fold extension

$$0 \to I^{(1)} \to S^p \xrightarrow{\alpha} \Gamma^p \to I^{(1)} \to 0$$

of [11, Lemma 4.12] where $\alpha_V : S^p(V) \to \Gamma^p(V)$ is the symmetrization homomorphism, $\alpha_V(v_1 \cdots v_p) = \sum_{\sigma \in \mathfrak{S}_p} v_{\sigma^{-1}(1)} \otimes \cdots \otimes v_{\sigma^{-1}(p)}$. It represents a nontrivial class by [11, Lemma 4.12, Theorem 1.2]. So

$$\cdots \to 0 \to S^p \xrightarrow{\alpha} \Gamma^p \to 0 \to \cdots$$

is not formal. See also Exercise 5.11.

Now let $A = I^{(r)}$. Then we write $\mathbf{K}^{\mathbf{r}}_A$ as $\mathbf{K}^{\mathbf{r}}$. Let E_r be the graded vector space of dimension p^r which equals k in dimensions $2i$, $0 \le i < p^r$. We view any graded vector space also as a cochain complex with zero differential and as a $\mathbb{G}_m$-module with weight j in degree j. For example, if $G \in \mathcal{P}_d$ then the $\mathbb{G}_m$ action on E_r induces one on G_{E_r} so that G_{E_r} is graded and thus a complex with differential zero.

A conjecture of Chałupnik, now says

Conjecture 5.7 (Formality). *For G in $\mathcal{P}_d$ one has $\mathbf{K}^{\mathbf{r}}(G^{(r)}) \cong G_{E_r}$ in $\mathcal{D}^b\mathcal{P}_d$. In particular, $\mathbf{K}^{\mathbf{r}}(G^{(r)})$ is formal.*

This is a variant of the collapsing conjecture of Touzé [32, Conjecture 8.1]. It is stronger. Both conjectures are theorems now [6], [33], 5.15.

Note that G_{E_r} is formal by definition. Note also that all its weights are even. So as a cochain complex it lives in even degrees. That already implies formality.

Exercise 5.8 (Compare [33, Corollary 5]). Let F, $G \in \mathcal{P}_d$. Assuming the formality conjecture, define a grading on $\mathrm{Ext}^\bullet(F, G_{E_r})$ so that its degree i subspace is isomorphic to $\mathrm{Ext}^i_{\mathcal{P}_{p^r d}}(F^{(r)}, G^{(r)})$.

Remark 5.9. Let $J^\bullet$ be a bounded injective coresolution of $G^{(r)}$. If one can show formality of $\mathbf{K^r}(J^\bullet)$, then one can also show it is quasi-isomorphic to G_{E_r}. The main problem is formality of $\mathbf{K^r}(J^\bullet)$. This problem does not require derived categories. But the problem is easier to motivate in the language of derived categories.

Following suggestions by Touzé let us give some evidence for the formality conjecture in the simplest case: $p = 2$, $r = 1$. Instead of $\mathbf{K^r}(G^{(1)})$ we will study $\mathbf{K^r}(G^{(1)}) \circ I^{(1)}$ and show that it is formal. So we will be off by one Frobenius twist. While we know how to untwist in $\mathcal{P}_d$ context, something more will be needed to do untwisting in $\mathcal{D}^b\mathcal{P}_d$ context. We postpone this issue until 5.15.

Now $\mathbf{K^r}(G^{(1)}) \circ I^{(1)}$ is represented by the complex $\mathrm{Hom}_{\mathcal{P}_{2d}}(\Gamma^{dV^{(1)}} \circ I^{(1)}, J^\bullet)$ in $\mathcal{P}_{2d}$, where $J^\bullet$ is a bounded injective coresolution of $G^{(1)}$. Observe that $\Gamma^{dV^{(1)}} \circ I^{(1)} = (\Gamma^{d(1)})^V$. This is where the extra twist helps: It turns out that $(\Gamma^{d(1)})^V$ is easier than $\Gamma^{dV} \circ I^{(1)}$. Rewrite our complex as $\mathrm{Hom}_{\mathcal{P}_{2d}}(J^{\bullet\#}, (S^{d(1)})_V)$. We first recall a standard injective coresolution of $(S^{d(1)})_V$.

5.10. A standard coresolution in characteristic two

It is here that the assumptions on p and r help. In general one needs the Troesch complexes to see that $\mathbf{K^r}(G^{(r)}) \circ I^{(r)}$ equals $G_{E_r} \circ I^{(r)}$ in $\mathcal{D}^b\mathcal{P}_{p^r d}$ and we refer to [32], [33] for details.

Let T be the group scheme of diagonal matrices in GL_2. If $W \in \mathcal{V}_k$ then T acts through k^2 on the symmetric algebra $S^*(k^2 \otimes_k W)$ with weight space $S^i(W) \otimes_k S^j(W)$ of weight (i,j). So the $S^i(W) \otimes_k S^j(W)$ are direct summands of $S^{i+j}_{k^2}(W)$ and $S^i \otimes_k S^j$ is an injective in $\mathcal{P}_{i+j}$ because it is a summand of an injective. Now recall $p = 2$. We make the algebra $S^*(k^2 \otimes_k W) = S^*W \otimes_k S^*W$ into a differential graded algebra with differential d whose restriction to $S^1(k^2 \otimes_k W)$ is given by $\begin{pmatrix} 0 & 0 \\ 1 & 0 \end{pmatrix} \otimes \mathrm{id}_W$. So if W has dimension one then the differential graded algebra is isomorphic to the polynomial ring $k[x, y]$ in two variables and the differential is $y\frac{\partial}{\partial x}$. The subcomplex $S^{2n}W \to S^{2n-1}W \otimes_k S^1W \to \cdots \to S^{2n-i}W \otimes_k S^iW \to \cdots \to S^1W \otimes_k S^{2n-1}W \to S^{2n}W$ is a coresolution of $S^nW^{(1)}$. (Exercise. Use the exponential property.) So we have a standard coresolution

$$S^{2n} \to S^{2n-1} \otimes_k S^1 \to \cdots \to S^{2n-i} \otimes_k S^i \to \cdots S^1 \otimes S^{2n-1} \to S^{2n} \to 0$$

of $S^{n(1)}$. We now coresolve $(S^{d(1)})_V$ by

$$R_V^{2d\bullet}: \quad S_V^{2d} \to S_V^{2d-1} \otimes_k S_V^1 \to \cdots S_V^{2d-i} \otimes_k S_V^i \to \cdots \to S_V^1 \otimes_k S_V^{2d-1} \to S_V^{2d}.$$

Exercise 5.11. Keep $p = 2$. Determine $\mathrm{Hom}_{\mathcal{P}_2}(S^2, S^2)$ and $\mathrm{Hom}_{\mathcal{P}_2}(\Gamma^2, I \otimes I)$. Show there is no nonzero cochain map from the complex

$$C^\bullet : \cdots \to 0 \to S^2 \xrightarrow{\alpha} \Gamma^2 \longrightarrow 0 \to 0 \to \cdots$$

of Remark 5.6 to the injective coresolution

$$D^\bullet : \cdots \to 0 \to S^2 \to I \otimes I \to S^2 \to 0 \to \cdots$$

of $I^{(1)}$. (In both complexes the first S^2 is placed in degree 0.) By [37, Cor 10.4.7] this implies there is no nonzero morphism from $C^\bullet$ to $D^\bullet$ in the derived category $\mathcal{D}^+\mathcal{P}_2$. Confirm the claim in Remark 5.6 that $C^\bullet$ is not formal.

5.12. Formality continued

We are studying the complex $\mathrm{Hom}_{\mathcal{P}_{2d}}(J^{\bullet\#}, (S^{d(1)})_V)$ in $\mathcal{P}_{2d}$ up to quasi-isomorphism. We may replace it with the total complex of the double complex $\mathrm{Hom}_{\mathcal{P}_{2d}}(J^{\bullet\#}, R_V^{2d\bullet})$ and then (by 'balance' [37, 2.7]) with the complex

$$\mathrm{Hom}_{\mathcal{P}_{2d}}(G^{(1)\#}, R_V^{2d\bullet}) \text{ in } \mathcal{P}_{2d}.$$

If one forgets the differential then this is just $\mathrm{Hom}_{\mathcal{P}_{2d}}(G^{(1)\#}, S^{2d}_{k^2\otimes_k V}) = G^{(1)}(k^2 \otimes_k V)$ and we now inspect its weight spaces for our torus T. Because of the Frobenius twist in $G^{(1)}$ the weights are all multiples of p, and p equals 2 now. On the other hand, on $\mathrm{Hom}_{\mathcal{P}_{2d}}(G^{(1)\#}, S_V^{2d-i} \otimes_k S_V^i)$ the weight is simply $(2d - i, i)$. So the only nonzero terms in the complex $\mathrm{Hom}_{\mathcal{P}_{2d}}(G^{(1)\#}, R_V^{2d\bullet})$ are in even degrees and formality follows. Moreover, in even degree $2i$ one gets the weight space of degree $(2d - 2i, 2i)$ of $G^{(1)}(k^2 \otimes_k V)$. Let $\mathbb{G}_m$ act on k^2 with weight zero on $(1, 0)$ and weight one on $(0, 1)$. So now $E_1 = (k^2)^{(1)}$ as $\mathbb{G}_m$-modules. As a (graded) functor in V we get that $\mathrm{Hom}_{\mathcal{P}_{2d}}(G^{(1)\#}, R_V^{2d\bullet})$ is $(G^{(1)})_{k^2}$ or $G_{E_1} \circ I^{(1)}$. So we have seen that for $p = 2$, $r = 1$ the complex $\mathbf{K}^\mathbf{r}(J^\bullet) \circ I^{(1)}$ is quasi-isomorphic to $G_{E_r} \circ I^{(1)}$. That means that $\mathbf{K}^\mathbf{r}(G^{(r)}) \circ I^{(1)} \cong G_{E_r} \circ I^{(1)}$ in $\mathcal{D}^+\mathcal{P}_{2d}$ for $p = 2$, $r = 1$.

Now we would like to untwist to get the formality conjecture for $p = 2$, $r = 1$. It is not obvious how to do that. One needs constructions with better control of the functorial behavior. In his solution in [30] of the (CFG) conjecture Touzé faced similar difficulties. As standard coresolutions he used Troesch coresolutions. They are not functorial. This is the main obstacle that he had to get around in order to construct the classes $c[m]$. His approach in [30] is to invent a new category, the twist-compatible category, on which the Troesch construction is functorial and which is just big enough to contain a repeated reduced bar construction that coresolves divided powers.

In the proof [33] of the collapsing conjecture a different argument is used. We call it untwisting the collapse of a hyper Ext spectral sequence. It comes next.

5.13. Untwisting the collapse of a hyper Ext spectral sequence

Let $C^\bullet$ be a bounded above complex in $\mathcal{P}_d$ and $D^\bullet$ a bounded below complex in $\mathcal{P}_d$. Put $C_i = C^{-i}$. Let $J^\bullet$ be a bounded below complex of injectives that coresolves $D^\bullet$. The homology groups of the total complex $\mathrm{Tot}\,\mathrm{Hom}_{\mathcal{P}_d}(C_\bullet, J^\bullet)$ of the bicomplex $\mathrm{Hom}_{\mathcal{P}_d}(C_i, J^j)$ are known as *hyper Ext groups* of $C^\bullet$, $D^\bullet$. Consider the second spectral sequence of the bicomplex $\mathrm{Hom}_{\mathcal{P}_d}(C_i, J^j)$

$$E_2^{ij} = H^i \,\mathrm{Hom}_{\mathcal{P}_d}(H_j(C_\bullet), J^\bullet) \Rightarrow H^{i+j}\,\mathrm{Tot}\,\mathrm{Hom}_{\mathcal{P}_d}(C_\bullet, J^\bullet).$$

We call it the *hyper Ext spectral sequence* associated with $(C^\bullet, D^\bullet)$. It is covariantly functorial in $D^\bullet$ and contravariantly functorial in $C^\bullet$. Say $\tilde{C}^\bullet \to C^\bullet$, $D^\bullet \to \tilde{D}^\bullet$ are quasi-isomorphisms. Let $\tilde{J}^\bullet$ be an injective coresolution of $\tilde{D}^\bullet$. Then $J^\bullet$, $\tilde{J}^\bullet$ are quasi-isomorphic complexes of injectives, hence homotopy equivalent. The hyper Ext spectral sequence associated with $(C^\bullet, D^\bullet)$ is isomorphic with the hyper Ext spectral sequence associated with $(\tilde{C}^\bullet, \tilde{D}^\bullet)$. (Check this.) In particular, if $C^\bullet$ is formal, then the spectral sequence is a direct sum of spectral sequences with just one row, so that the spectral sequence degenerates at page two. We also say that the spectral sequence *collapses*.

Now suppose that we do not know that $C^\bullet$ is formal, but only that $C^{\bullet(1)}$ is formal. Frobenius twist $G \mapsto G^{(1)}$ defines an embedding of $\mathcal{P}_d$ into $\mathcal{P}_{dp}$. Coresolving $J^{\bullet(1)}$ we get a map from the hyper Ext spectral sequence E of $(C^\bullet, D^\bullet)$ to the hyper Ext spectral sequence $\tilde{E}$ of $(C^{\bullet(1)}, D^{\bullet(1)})$. Now we make the extra assumption that $D^\bullet$ is concentrated in one degree. Say degree zero, to keep notations simple. Write $D^\bullet$ as D. Then the second page of E is given by $E_2^{ij} = \mathrm{Ext}^i_{\mathcal{P}_d}(H_j(C_\bullet), D)$ and the second page of $\tilde{E}$ is given by $\tilde{E}_2^{ij} = \mathrm{Ext}^i_{\mathcal{P}_{dp}}(H_j(C_\bullet)^{(1)}, D^{(1)})$. Now the map $E_2^{ij} \to \tilde{E}_2^{ij}$ is injective by the Twist Injectivity Theorem 4.9. We conclude that E itself degenerates at page two by means of the following basic lemma about spectral sequences.

Lemma 5.14. *Let $E \to \tilde{E}$ be a morphism of spectral sequences that is injective at the second page. If $\tilde{E}$ degenerates at page two, then so does E.*

Proof. The second page E_2 of E with differential d_2 may be viewed as a subcomplex of $\tilde{E}_2$ with differential $\tilde{d}_2$. So the differential d_2 of E_2 vanishes and $E_3 \to \tilde{E}_3$ is also injective. But the differential of $\tilde{E}_3$ vanishes, so the differential of E_3 vanishes again. Repeat. □

So we do *not* need the formality of $C^\bullet$ to conclude the collapsing of $E_2^{ij} = \mathrm{Ext}^i_{\mathcal{P}_d}(H_j(C_\bullet), D) \Rightarrow H^{i+j}\,\mathrm{Tot}\,\mathrm{Hom}_{\mathcal{P}_d}(C_\bullet, J^\bullet)$. Formality of $C^{\bullet(1)}$ suffices. We have 'untwisted' the collapsing.

Exercise 5.15 (Untwisting Formality). Let $C^\bullet$ be a bounded complex in $\mathcal{P}_d$ such that $C^{\bullet(1)}$ is formal and let $D^\bullet$ a formal bounded below complex in $\mathcal{P}_d$. Show that the hyper Ext spectral sequence E of $(C^\bullet, D^\bullet)$ collapses. Say $C^\bullet$ has nonzero cohomology. Put $m = \min\{\, i \mid H^i(C^\bullet) \neq 0 \,\}$. Now take for $D^\bullet$ the maximal subcomplex of $C^\bullet$ with $D^i = 0$ for $i > m$. Then $H^m(D^\bullet) \to H^m(C^\bullet)$ is an isomorphism and $H^i(D^\bullet) = 0$ for $i \neq m$. The complex $D^\bullet$ is formal. Let $D^\bullet \to J^\bullet$ be an injective coresolution again. Recall that $H^0(\mathrm{Tot}\,\mathrm{Hom}_{\mathcal{P}_d}(C_\bullet, J^\bullet)) = \mathrm{Hom}_{\mathcal{K}^+\mathcal{P}_d}(C^\bullet, J^\bullet)$ by [37, 2.7.5]. Compare the collapsed hyper Ext spectral sequences of $(C^\bullet, D^\bullet)$ and $(D^\bullet, D^\bullet)$. Show that $\mathrm{Hom}_{\mathcal{K}^+\mathcal{P}_d}(C^\bullet, J^\bullet) \to \mathrm{Hom}_{\mathcal{K}^+\mathcal{P}_d}(D^\bullet, J^\bullet)$ is surjective. Choose

$f : C^\bullet \to J^\bullet$ so that the composite $D^\bullet \to C^\bullet \xrightarrow{f} J^\bullet$ is a quasi-isomorphism. Show that $C^\bullet$ is quasi-isomorphic to $J^\bullet \oplus (C^\bullet/D^\bullet)$. Show that $C^\bullet$ is formal by induction on the number of nontrivial $H^i(C^\bullet)$. This establishes untwisting of formality. Notice that we did not mention $\mathcal{D}^b\mathcal{P}_d$ in this exercise. But recall that $\mathrm{Hom}_{\mathcal{D}^+\mathcal{P}_d}(C^\bullet, D^\bullet)$ may be identified with $\mathrm{Hom}_{\mathcal{K}^+\mathcal{P}_d}(C^\bullet, J^\bullet)$.

This finishes the proof of the formality conjecture for the case: $p = 2$, $r = 1$.

6. Bifunctors and CFG

There are some more ingredients entering into the proof of the CFG theorem in [31]. The paper [31] has an extensive introduction, which we recommend to the reader. We now provide a companion to that introduction.

The proof of the CFG conjecture takes several steps. First one reduces to the case of GL_n. This uses a transfer principle, reminiscent of Shapiro's Lemma, that can be traced back to the nineteenth century. Next one needs to know about Grosshans graded algebras and good filtrations. The case were the coefficient algebra A is a Grosshans graded algebra lies in between the general case and the case of good filtration. In the good filtration case CFG is known by invariant theory. There is a spectral sequence connecting the Grosshans graded case with the general case and another spectral sequence connecting it with the good filtration case. We need to get these spectral sequences under control. That is done by finding an algebra of operators, operating on the spectral sequences, and establishing finiteness properties of the spectral sequences with respect to the operators. It is here that the classes of Touzé come in. They allow a better grip on the operators.

Now we introduce some of these notions.

6.1. Costandard modules

Let $k = \mathbb{F}_p$ and put $G = GL_n$, $n \geq 2$. We have already introduced the torus T of diagonal matrices. Our standard Borel group B will be the subgroup scheme with $B(R)$ equal to the subgroup of upper triangular matrices of $G(R)$. Similarly U, the unipotent radical of B, is the subgroup scheme with $U(R)$ equal to the subgroup of upper triangular matrices with ones on the diagonal. The *Grosshans height* ht, also known as the sum of the coroots associated to the positive roots, is given by

$$\mathrm{ht}(\lambda) = \sum_{i<j} \lambda_i - \lambda_j = \sum_i (n - 2i + 1)\lambda_i.$$

Here we use the ancient convention that the roots of B are positive. If V is a representation of G, let us say that it has *highest weight* λ if λ is a weight of V and all other weights μ have strictly smaller Grosshans height $\mathrm{ht}(\mu)$. (This nonstandard convention is good enough for the present purpose.) Irreducible G-modules have a highest weight and are classified up to isomorphism by that weight. Write $L(\lambda)$ for the irreducible module with highest weight λ. The weight space of weight λ in $L(\lambda)$ is one-dimensional and equal to the subspace $L(\lambda)^U$ of U-invariants.

We now switch to geometric language as if we are dealing with varieties. In other words, we switch from the setting of group schemes [7], [18], [36] to algebraic groups and varieties defined over $\mathbb{F}_p$ [29].

The *flag variety* [29, 8.5] G/B is a *projective variety* [17, I §2], [29, 1.7], not an affine variety. Given $L(\lambda)$ as above there is an *equivariant line bundle* [29, 8.5.7] $\mathcal{L}_\lambda$ on G/B so that its module $\nabla(\lambda)$ of global *sections* [29, 8.5.7–8] on G/B has a unique irreducible submodule, and this submodule is isomorphic to $L(\lambda)$. The *costandard module* $\nabla(\lambda)$ is finite-dimensional (because G/B is a projective variety). The weight space of weight λ in $\nabla(\lambda)$ is also one-dimensional and equal to the subspace $\nabla(\lambda)^U$ of U-invariants. Every other G-module V whose weight space V_λ of weight λ is one-dimensional and equal to V^U embeds into $\nabla(\lambda)$. *Kempf vanishing* [18, II Chapter 4] says that $\mathcal{L}_\lambda$ has no higher sheaf cohomology on G/B. One derives from this [5] that $H^{>0}(G, \nabla(\lambda))$ vanishes. All nontrivial cohomology of G-modules is due to the distinction between the irreducible modules $L(\lambda)$ and the costandard modules $\nabla(\lambda)$. The dimensions of the weight spaces of $\nabla(\lambda)$ are given by the famous Weyl character formula

$$\mathrm{Char}(\nabla(\lambda)) = \frac{\sum_{w\in W}(-1)^{\ell(w)}e^{w(\lambda+\rho)}}{e^\rho\prod_{\alpha>0}(1-e^\alpha)}.$$

We do not explain the precise meaning here but just observe that the formula is characteristic free. The dimensions of the weight spaces of $\nabla(\lambda)$ are the same as in the irreducible $GL_n(\mathbb{C})$-module with highest weight λ. Determining the dimensions of the weight spaces of $L(\lambda)$ is less easy in general, to put it mildly.

Example 6.2. Let $V = k^n$ be the defining representation of GL_n over $\mathbb{F}_p$. The symmetric powers $S^m(V)$ are costandard modules. More specifically, $S^m(V)$ is $\nabla((m,0,\ldots,0))$. When $m = p^r$ the irreducible submodule $L((m,0,\ldots,0))$ of $S^m(V)$ is spanned by the v^{p^r}.

If V is a nonzero G-submodule of $\nabla(\lambda)$ then it determines a map ϕ_V from the flag variety G/B to the projective space whose points are codimension one subspaces of V, or one-dimensional subspaces of $V^\vee$. (To a point of G/B one associates the codimension one subspace of V consisting of sections vanishing at the point. Then one takes the elements in the dual that vanish on the codimension one subspace.) The image of G/B under ϕ_V is isomorphic to $G/\tilde{P}$, where $\tilde{P}$ is the scheme theoretic stabilizer of the image of the point B. Here 'scheme theoretic' indicates that the functorial interpretation of group schemes is needed. The image of the point B is the highest weight space of $V^\vee$. The group scheme $\tilde{P}$ need not be reduced [21], [38], but the image of $\tilde{P}$ under a sufficiently high power F^r of the Frobenius homomorphism $F : G \to G$ is the stabilizer P of the highest weight space of $\nabla(\lambda)^\vee$. This P is an ordinary *parabolic subgroup* [29, 6.2] and thus reduced, meaning that its coordinate ring is reduced. There is a graded algebra associated with the image of ϕ_V. This algebra A_V is known as *coordinate ring of the affine cone* over the image of ϕ_V. It is a graded k-algebra, generated as a k-algebra by its degree one part, which is V. This is typical for closed subsets of a projective space: Such a subset does not have an ordinary coordinate ring like an affine variety would, but a graded coordinate ring [17, II Corollary 5.16].

Similarly one has a graded algebra

$$A_{\nabla(\lambda)} = \bigoplus_{m\geq 0} \Gamma(G/B, \mathcal{L}_\lambda^m)$$

associated with the image G/P of G/B in the projective space whose points are codimension one subspaces of $\nabla(\lambda)$. The algebra A_V may be embedded into $A_{\nabla(\lambda)}$. Mathieu observed [20, 3.4] that the two affine cones have the same rational points over fields and concluded from this that for $r \gg 0$ the smaller algebra contains all f^{p^r} for f in the larger algebra. This is not always the same r as in F^r above.

6.3. Grosshans filtration

The situation above generalizes. If V is a possibly infinite-dimensional G-module we define its *Grosshans filtration* to be the filtration $V_{\leq -1} = 0 \subseteq V_{\leq 0} \subseteq V_{\leq 1} \subseteq V_{\leq 2} \cdots$ where $V_{\leq i}$ is the largest G-submodule of V all whose weights μ satisfy $\mathrm{ht}(\mu) \leq i$. The associated graded $\bigoplus_i V_{\leq i}/V_{\leq i-1}$ we call the *Grosshans graded* $\mathrm{gr}\, V$. It can naturally be embedded into a direct sum $\mathrm{hull}_\nabla(\mathrm{gr}\, V)$ of costandard modules in such a way that no new U-invariants are introduced: $(\mathrm{gr}\, V)^U = (\mathrm{hull}_\nabla(\mathrm{gr}\, V))^U$. We say that V has *good filtration* [18, II 4.16 Remarks] if $\mathrm{gr}\, V$ itself is a direct sum of costandard modules, in which case $\mathrm{gr}\, V = \mathrm{hull}_\nabla(\mathrm{gr}\, V)$ [14, Theorem 16]. As costandard modules have no higher G-cohomology, a module with good filtration has vanishing higher G-cohomology. One says that a module has *finite good filtration dimension* if it has a finite coresolution by modules with good filtration. Such a module has only finitely many nonzero G-cohomology groups.

If $A \in \mathsf{Rg}_k$ is a k-algebra with G-action, so that the multiplication map $A \otimes_k A \to A$ is a G-module map, then $\mathrm{gr}\, A$ and $\mathrm{hull}_\nabla(grA)$ are also k-algebras with G-action. Moreover, if A is of finite type, then so are $\mathrm{gr}\, A$ and $\mathrm{hull}_\nabla(grA)$ by Grosshans [14]. And then there is an r so that $\mathrm{gr}\, A$ contains all f^{p^r} for f in the larger algebra $\mathrm{hull}_\nabla(grA)$. All higher G-cohomology of A is due to the distinction between $\mathrm{gr}\, A$ and $\mathrm{hull}_\nabla(\mathrm{gr}\, A)$. It is here that Frobenius twists and Frobenius kernels enter the picture. (In this subject area a *Frobenius kernel* refers to the finite group scheme which is the scheme theoretic kernel of an iterated Frobenius map $F^r : G \to G$.) In general we have no grip on the size of the minimal r so that $\mathrm{gr}\, A$ contains all f^{p^r}. This is where the results get much more qualitative than those of Friedlander and Suslin.

Problem 6.4. *Given your favorite A, estimate the r such that* $\mathrm{gr}\, A$ *contains all f^{p^r} for f in the larger algebra* $\mathrm{hull}_\nabla(\mathrm{gr}\, A)$. *Such an estimate is desirable because one may give a bound on the Krull dimension of $H^{\mathrm{even}}(G, A)$ in terms of r, n and* $\dim A$ *by inspecting the proof in* [31].

6.5. The classes of Touzé

The *adjoint representation* $\mathfrak{gl}_n$ of GL_n is defined as the k-module of n by n matrices over k with $GL_n(R)$ acting by conjugation on the set $M_n(R) = \mathfrak{gl}_n \otimes_k R$ of n by n

matrices over R. This is also known as the adjoint action on the Lie algebra. The adjoint representation is not a polynomial representation as soon as $n \geq 2$. We now have all the ingredients to state the theorem of Touzé on *lifted classes* proved using strict polynomial bifunctors. The base ring is our field $k = \mathbb{F}_p$ and $n \geq 2$.

Theorem 6.6 (Touzé [30]. Lifted universal cohomology classes). *There are cohomology classes $c[m]$ so that*

1. $c[1] \in H^2(GL_n, \mathfrak{gl}_n^{(1)})$ *is nonzero,*
2. *For $m \geq 1$ the class $c[m] \in H^{2m}(GL_n, \Gamma^m(\mathfrak{gl}_n^{(1)}))$ lifts $c[1] \cup \cdots \cup c[1] \in H^{2m}(GL_n, \bigotimes^m(\mathfrak{gl}_n^{(1)}))$.*

6.7. Strict polynomial bifunctors

The representations $\Gamma^m(\mathfrak{gl}_n^{(1)}))$ in the 'lifted classes' theorem of Touzé are not polynomial. To capture their behavior one needs the *strict polynomial bifunctors* of Franjou and Friedlander [10]. We already encountered them in disguise when discussing parametrized functors. An example of a strict polynomial bifunctor is the bifunctor

$$\mathrm{Hom}_{\Gamma^d\mathcal{V}_k}(-_1, -_2) : \Gamma^d\mathcal{V}_k^{\mathrm{op}} \times \Gamma^d\mathcal{V}_k \to \mathcal{V}_k;\ (V, W) \mapsto \mathrm{Hom}_{\Gamma^d\mathcal{V}_k}(V, W).$$

It is contravariant in $-_1$ and covariant in $-_2$. More generally one could consider the category $\mathcal{P}_e^d$ of k-bilinear functors $\Gamma^d\mathcal{V}_k^{\mathrm{op}} \times \Gamma^e\mathcal{V}_k \to \mathcal{V}_k$. Do not get confused by the strange notation $\mathcal{P}^{\mathrm{op}} \times \mathcal{P}$ for $\bigoplus_{d,e} \mathcal{P}_e^d$ used in [10]. It is not a product.

If $\mathcal{A}$ and $\mathcal{B}$ are k-linear categories, then one can form the k-linear category $\mathcal{A} \otimes_k \mathcal{B}$ whose objects are pairs (A, B) with $A \in \mathcal{A}$, $B \in \mathcal{B}$. For morphisms one puts $\mathrm{Hom}_{\mathcal{A}\otimes_k\mathcal{B}}((A, B), (A', B')) = \mathrm{Hom}_{\mathcal{A}}(A, A') \otimes_k \mathrm{Hom}_{\mathcal{B}}(B, B')$. One may then define the category $\mathcal{P}_e^d$ of strict polynomial bifunctors of bidegree (d, e) to be the category of k-linear functors from $\Gamma^d\mathcal{V}_k^{\mathrm{op}} \otimes_k \Gamma^e\mathcal{V}_k$ to $\mathcal{V}_k$.

One gets more bifunctors by composition. For instance, $\Gamma^m(\mathfrak{gl}^{(1)})$ is the strict polynomial bifunctor of bidegree (mp, mp) sending (V, W) to $\Gamma^m(\mathfrak{gl}(V^{(1)}, W^{(1)}))$, where $\mathfrak{gl}$ means Hom_k. The GL_n-module $\Gamma^m(\mathfrak{gl}_n^{(1)})$ is obtained by substituting k^n for both V and W in $(V, W) \mapsto \Gamma^m(\mathfrak{gl}^{(1)})(V, W)$. Such substitution defines a functor $\mathcal{P}_d^d \to \mathsf{Mod}_{GL_n}$ and for $n \geq d$ a theorem of Franjou and Friedlander gives

$$\mathrm{Ext}^\bullet_{\mathcal{P}_d^d}(\Gamma^d \mathfrak{gl}, F) \cong H^\bullet(GL_n, F(k^n, k^n)).$$

The map from the left-hand side to the right-hand side goes by way of $\mathrm{Ext}^\bullet_{GL_n}(\Gamma^d \mathfrak{gl}_n, F(k^n, k^n))$. The invariant $\mathrm{id}^{\otimes d} \in \Gamma^d \mathfrak{gl}_n$ gives a GL_n-module map $k \to \Gamma^d \mathfrak{gl}_n$ which allows one to go on to

$$\mathrm{Ext}^\bullet_{GL_n}(k, F(k^n, k^n)) = H^\bullet(GL_n, F(k^n, k^n)).$$

6.8. The collapsing theorem for bifunctors

In order to explain the collapsing theorem for strict polynomial bifunctors in [33] we need to introduce a few counterparts of definitions given above for strict polynomial functors.

So let $B \in \mathcal{P}_e^d$ be a bifunctor and let $Z \in \Gamma^e\mathcal{V}$. Then the parametrized bifunctor B_Z is defined by parametrizing the covariant variable: $B_Z(V,W) = B(V, Z \otimes_k^{\Gamma^d} W)$. Now assume Z comes with an action of $\mathbb{G}_m$. Then $\mathbb{G}_m$ acts on B_Z and we write B_Z^t for the weight space of weight t. The example we have in mind is $Z = E_r$ as in 5.4.

Define a Frobenius twist $B^{(r)}$ of B by precomposition with $I^{(r)}$ in both variables: $B^{(r)}(V,W) = B(V^{(r)}, W^{(r)})$.

Theorem 6.9 (Collapsing of the Twisting Spectral Sequence [33]). *Let $B \in \mathcal{P}_d^d$. There is a first quadrant spectral sequence*

$$E_2^{st} = \mathrm{Ext}^s_{\mathcal{P}_d^d}(\Gamma^d \mathfrak{gl}, B^t_{E_r}) \Rightarrow \mathrm{Ext}^{s+t}_{\mathcal{P}_{pd}^{pd}}(\Gamma^{dp} \mathfrak{gl}, B^{(r)})$$

and this spectral sequence collapses.

The proof of this theorem uses the themes that we have seen above for ordinary polynomial functors:

- adjoint of the twist,
- formality after twisting,
- untwisting a collapse.

Once one has the theorem one gets a much better grip on the connection between $H^{2m}(GL_n, \Gamma^m(\mathfrak{gl}_n^{(1)}))$ and $H^{2m}(GL_n, \bigotimes^m(\mathfrak{gl}_n^{(1)}))$. This then leads to the second generation construction of the classes of Touzé [33].

6.10. How the classes of Touzé help

We find it hard to improve on the introduction to [31]. Go read it.

References

[1] D.J. Benson, *Representations and cohomology. I. Basic representation theory of finite groups and associative algebras.* Second edition. Cambridge Studies in Advanced Mathematics, 30. Cambridge University Press, Cambridge, 1998.

[2] D.J. Benson, *Representations and cohomology. II. Cohomology of groups and modules.* Second edition. Cambridge Studies in Advanced Mathematics, 31. Cambridge University Press, Cambridge, 1998.

[3] A. Borel, *Essays in the history of Lie groups and algebraic groups.* History of Mathematics, 21. American Mathematical Society, Providence, RI; London Mathematical Society, Cambridge, 2001.

[4] K.S. Brown, *Cohomology of groups*, Graduate Texts in Mathematics 87, Springer-Verlag 1982.

[5] E. Cline, B. Parshall, L. Scott, W. van der Kallen, *Rational and generic cohomology,* Invent. Math. 39 (1977), 143–163.

[6] M. Chałupnik, *Derived Kan extension for strict polynomial functors,* `arXiv:1106.3362v2`, Int. Math. Res. Notices, online first January 16, 2015, doi:10.1093/imrn/rnu269

[7] M. Demazure, P. Gabriel, *Groupes algébriques.* Tome I. Masson & Cie, Paris; North-Holland Publishing Co., Amsterdam, 1970.

[8] D. Eisenbud, *Commutative algebra. With a view toward algebraic geometry.* Graduate Texts in Mathematics, 150. Springer-Verlag, New York, 1995.

[9] L. Evens, *The cohomology ring of a finite group*, Trans. Amer. Math. Soc. 101 (1961), 224–23.

[10] V. Franjou, E.M. Friedlander, *Cohomology of bifunctors*, Proceedings of the London Mathematical Society 97 (2008), 514–544. doi:10.1112/plms/pdn005

[11] E.M. Friedlander, A.A. Suslin, *Cohomology of finite group schemes over a field*, Invent. Math. 127 (1997), 209–270.

[12] V. Franjou and W. van der Kallen, *Power reductivity over an arbitrary base*, Documenta Mathematica, Extra Volume Suslin (2010), 171–195.

[13] P. Gordan, *Beweis dass jede Covariante und Invariante einer binären Form eine ganze Function mit numerischen Coefficienten einer endlichen Anzahl solcher Formen ist*, J.f.d. reine u. angew. Math. 69 (1868), 323–354.

[14] F.D. Grosshans, *Contractions of the actions of reductive algebraic groups in arbitrary characteristic*, Invent. Math. 107 (1992), 127–133.

[15] F.D. Grosshans, *Algebraic homogeneous spaces and invariant theory*, Lecture Notes in Mathematics, 1673. Springer-Verlag, Berlin, 1997.

[16] W.J. Haboush, *Reductive groups are geometrically reductive.* Ann. of Math. (2) 102 (1975), 67–83.

[17] R. Hartshorne, *Algebraic Geometry.* Graduate Texts in Mathematics, Springer, Berlin, 1977.

[18] J.-C. Jantzen, *Representations of Algebraic Groups*, Mathematical Surveys and Monographs vol. 107, Amer. Math. Soc., Providence, 2003.

[19] S. Mac Lane, *Categories for the Working Mathematician*, Graduate Texts in Mathematics, 5, Springer-Verlag, 1998, ISBN 0-387-98403-8

[20] O. Mathieu, *Filtrations of G-modules*, Ann. Sci. École Norm. Sup. 23 (1990), 625–644.

[21] N. Lauritzen, *Embeddings of homogeneous spaces in prime characteristics.* Amer. J. Math. 118 (1996), 377–387.

[22] D. Mumford, *Geometric Invariant Theory.* Ergebnisse der Mathematik und ihrer Grenzgebiete, Neue Folge, Band 34. Springer-Verlag, Berlin-New York, 1965

[23] M. Nagata, *Invariants of a group in an affine ring*, J. Math. Kyoto Univ. 3 (1963/1964), 369–377.

[24] E. Noether, *Der Endlichkeitssatz der Invarianten endlicher linearer Gruppen der Charakteristik p*, Nachr. Ges. Wiss. Göttingen (1926), 28–35.

[25] T. Pirashvili, *Introduction to functor homology.* Rational representations, the Steenrod algebra and functor homology, 1–26, Panor. Synthèses, 16, Soc. Math. France, Paris, 2003.

[26] T. Pirashvili, *Polynomial functors over finite fields* (*after Franjou, Friedlander, Henn, Lannes, Schwartz, Suslin*). Séminaire Bourbaki, Vol. 1999/2000. Astérisque 276 (2002), 369–388.

[27] N. Roby, *Construction de certaines algèbres à puissances divisées.* Bull. Soc. Math. France 96 (1968), 97–113.

[28] T.A. Springer, *Invariant theory.* Lecture Notes in Mathematics, 585. Springer-Verlag, Berlin-New York, 1977.

[29] T.A. Springer, *Linear algebraic groups.* Second edition. Progress in Mathematics, 9. Birkhäuser Boston, Inc., Boston, MA, 1998. ISBN: 0-8176-4021-5

[30] A. Touzé, *Universal classes for algebraic groups*, Duke Mathematical Journal, Vol. 151 (2010), 219–249. DOI: 10.1215/00127094-2009-064

[31] A. Touzé and W. van der Kallen, *Bifunctor cohomology and Cohomological finite generation for reductive groups*, Duke Mathematical Journal, Vol. 151 (2010), 251–278. DOI: 10.1215/00127094-2009-065

[32] A. Touzé, *Troesch complexes and extensions of strict polynomial functors.* Annales scientifiques de l'ENS 45, fascicule 1 (2012), 53–99.

[33] A. Touzé, *A construction of the universal classes for algebraic groups with the twisting spectral sequence*, Transformation Groups, Vol. 18 (2013), 539–556.

[34] W. van der Kallen, *Finite good filtration dimension for modules over an algebra with good filtration.* J. Pure Appl. Algebra 206 (2006), 59–65.

[35] W. van der Kallen, *Good Grosshans filtration in a family*, `arXiv:1109.5822`

[36] W.C. Waterhouse, *Introduction to affine group schemes.* Graduate Texts in Math. 66, Springer-Verlag, Berlin-New York, 1994.

[37] C.A. Weibel, *An introduction to homological algebra.* Cambridge Studies in Advanced Mathematics, 38, Cambridge University Press, Cambridge, 1994.

[38] C. Wenzel, *Classification of all parabolic subgroup-schemes of a reductive linear algebraic group over an algebraically closed field.* Trans. Amer. Math. Soc. 337 (1993), 211–218.

Wilberd van der Kallen
Mathematisch Instituut Universiteit Utrecht
P.O. Box 80.010
NL-3508 TA Utrecht, The Netherlands
e-mail: `W.vanderKallen@uu.nl`

Progress in Mathematics, Vol. 311, 67–98

Polynomial Functors and Homotopy Theory

Roman Mikhailov

Abstract. We develop methods for proving that certain extensions of polynomial functors do not split naturally. As an application we give a functorial description of the third and the fourth stable homotopy groups of the classifying spaces of free abelian groups.

Mathematics Subject Classification (2010). 18E25, 55P65.

Keywords. Derived functor, homotopy group, polynomial functor.

1. Introduction

Eilenberg–Mac Lane spaces $K(A, n)$ are examples of spaces whose homotopy type depends naturally on an abelian group A. Other meaningful examples can be built from these: suspensions $\Sigma^m K(A, n)$, wedges $K(A, n) \vee K(A, m)$, etc. The homology and homotopy groups of these spaces can be viewed as functors on the category of abelian groups. Describing these as functors is more difficult than giving an abstract description.

This difficulty can be understood with the case of the homology $H_*(A)$ of discrete abelian groups (with trivial integral coefficients). The abelian groups $H_n(A)$ are obtained from the computation of the homology of cyclic groups by the Künneth formula, whereas the functors $A \mapsto H_n(A)$ are not [6].

In practice, the functorial description of different homological or homotopical functors is often obtained with the help of spectral sequences, depending functorially on an abelian group A. For example, suppose that we want to determine the functor $A \mapsto \pi_*(\Sigma K(A, 1))$. There is a functorial spectral sequence (see Section 6 for the description of this spectral sequence):

$$E^1_{i,j}(A) \Rightarrow \pi_{j+1}(\Sigma K(A, 1)) .$$

This material is based upon work supported by the National Science Foundation under agreement No. DMS-0635607. Any opinions, findings and conclusions or recommendations expressed in this material are those of the authors and do not necessarily reflect the views of the National Science Foundation. This research is supported by JSC "Gazprom Neft", as well as by the RF Presidential grant MD-381.2014.1.

In this spectral sequence, the $E^i_{i,j}(A)$ are functors of abelian group A, the differentials of the spectral sequence are natural transformations of functors, and the convergence of the spectral sequence is functorial. Assume that we can compute the first page of such a functorial spectral sequence. Then we are left with two problems.

1. Compute the differentials in the spectral sequence.
2. Solve the extension problems. Indeed, the E_∞-term is only isomorphic to the graded associated to a filtration on the abutment. So we have to determine how to glue the different pieces together.

To compute the differentials, we have to study natural transformations between functors. Given functors F, G, there are much less natural transformations $F \to G$ than morphisms of abelian groups $F(A) \to G(A)$, so functoriality usually greatly helps in determining the differentials of spectral sequences. To solve the extension problems, we have to study extensions between functors. In this paper, we present some methods to determine Hom and Ext groups between functors, and we show how they can be successfully applied in computations. Note that all the functors considered in this paper are over $\mathbb{Z}$, they are defined on the category of (free) abelian groups; observe that analogous results over fields can be obtained more easily. This paper continues the research started in [6], [7].

To give some examples which illustrate the spirit of questions considered in this paper, let us start with two complexes of abelian groups C_*, D_*. One can compute the homology of their tensor product $H(C_* \otimes D_*)$ in terms of $H(C_*)$ and $H(D_*)$ using the well-known Künneth formula. Now consider three abelian groups A, B, C. The Künneth formula gives the following exact sequence

$$0 \to \operatorname{Tor}(A, B) \otimes C \to H_1\left(A \overset{L}{\otimes} B \overset{L}{\otimes} C\right) \to \operatorname{Tor}(A \otimes B, C) \to 0 \tag{1.1}$$

which splits as a sequence of abelian groups. The middle term of this sequence is the functor $\operatorname{Trip}(A, B, C)$ of Mac Lane [22] which is simply the first homology group of the derived iterated tensor product. S. Mac Lane proved [22] that the sequence (1.1) does not split naturally as a sequence of multi-functors. On the other hand, let us fix the two groups $B = C = \mathbb{Z}/2$. In this case the sequence (1.1) has the form

$$0 \to \operatorname{Tor}(A, \mathbb{Z}/2) \to H_1\left(A \overset{L}{\otimes} \mathbb{Z}/2 \overset{L}{\otimes} \mathbb{Z}/2\right) \to A \otimes \mathbb{Z}/2 \to 0,$$

and this sequence splits as a sequence of functors. This simply follows from the fact that we can choose a splitting $A \overset{L}{\otimes} \mathbb{Z}/2 \overset{L}{\otimes} \mathbb{Z}/2 \simeq A \overset{L}{\otimes} (\mathbb{Z}/2 \oplus \mathbb{Z}/2[1])$ in the derived category functorially in A[1].

One more example is the following. Let A be an abelian group. A description of the third homology of A as a functor is given in [6]. There is a natural exact

[1] We will always use the traditional notation $[n]$ for the shift of degree n in the derived category.

sequence

$$0 \to \Lambda^3(A) \to H_3(A) \to \Omega_2(A) \to 0 \tag{1.2}$$

where Λ^3 is the third exterior power and Ω_2 is the quadratic functor due to Eilenberg–Mac Lane, which is in fact the first derived functor of the exterior square. The sequence (1.2) splits as a sequence of abelian groups. One can ask weather (1.2) splits as a sequence of functors. We prove (see Corollary 3.1) that this is not the case. More generally, we will prove that, for all $n \geq 3$, the natural injection $\Lambda^n(A) \hookrightarrow H_n(A)$ induced by the Pontryagin product in homology, does not split naturally (see Proposition 3.1)[2].

On the other hand, there are some cases in which the existence of a functorial splitting can be proved. For an object C of the derived category of abelian groups concentrated in non-positive dimensions $\mathsf{DAb}_{\leq 0}$, we show that the exact triangle in $\mathsf{DAb}_{\leq 0}$

$$LS^2(C[1]) \to L\Gamma_2(C[1]) \to C \overset{L}{\otimes} \mathbb{Z}/2[1] \to LS^2(C[1])[1]$$

induces a functorial splitting (Theorem 4.1)

$$\pi_i(L\Gamma_2(C[1])) \simeq \pi_i(LS^2(C[1]))) \oplus \pi_i\left(C \overset{L}{\otimes} \mathbb{Z}/2[1]\right), \ i \geq 1. \tag{1.3}$$

While the well-known theorem of Dold [13] implies that there is a splitting on the level of complexes which induces the splitting (1.3) on homotopy, we show that this splitting is not functorial, i.e.,

$$L\Gamma_2(C[1]) \neq LS^2(C[1]) \oplus C \overset{L}{\otimes} \mathbb{Z}/2[1]$$

in the derived category $\mathsf{DAb}_{\leq 0}$.

The paper is organized as follows. We recall in Section 2 the description of the polynomial functors on the category of free abelian groups in terms of maps between cross-effects [4]. The language of polynomial $\mathbb{Z}$-modules developed in [4] and [2] is useful for the description of Hom and Ext-groups for polynomial functors in the category of free abelian groups. We use this language for proving that certain exact sequences do not split.

We describe the third and fourth stable homotopy group of $K(A, 1)$ as a functor in Section 5. We show that, for a free abelian group A, there is a short exact sequence

$$0 \to S^2(A) \otimes \mathbb{Z}/2 \to \pi_3^S K(A, 1) \to \Lambda^3(A) \to 0$$

which does not split naturally. Moreover, the functor $\pi_3^S K(A, 1)$ represents the unique non-trivial element in the group of functorial extensions $\mathrm{Ext}(\Lambda^3, S^2 \otimes \mathbb{Z}/2) = \mathbb{Z}/2$. Section 4 deals with derived functors. In Section 6, functorial information is used to compute spectral sequences and solve the extension problems to their abutment.

[2]After posting the paper the author was informed by N.Kuhn that this result follows from Example 7.6 [20].

2. Polynomial functors

Denote by Ab (resp. fAb) the category of finitely generated abelian (resp. f.g. free abelian) groups. For a small category C, let $Fun(\mathsf{C},\mathsf{Ab})$ be the category of functors from C to Ab. Morphisms in $Fun(\mathsf{C},\mathsf{Ab})$ are natural transformations between functors. It is well known that $Fun(\mathsf{C},\mathsf{Ab})$ is an abelian category with enough projectives and injectives. By $\mathsf{DAb}_{\leq 0}$ we mean the derived category of abelian groups living in non-negative degrees which is equivalent to the homotopy category of simplicial abelian groups via the Dold–Kan correspondence [14].

The main functors which we will consider are the following ($n \geq 1$):

- Tensor powers $\otimes^n : \mathsf{Ab} \to \mathsf{Ab}$
- Symmetric powers $S^n : \mathsf{Ab} \to \mathsf{Ab}$
- Exterior powers $\Lambda^n : \mathsf{Ab} \to \mathsf{Ab}$
- Divided powers $\Gamma_n : \mathsf{Ab} \to \mathsf{Ab}$
- Antisymmetric square $\tilde{\otimes}^2 : \mathsf{Ab} \to \mathsf{Ab}$, defined as

$$\tilde{\otimes}^2(A) := A \otimes A / \{a \otimes b + b \otimes a,\ a, b \in A\}.$$

The *nth divided power functor* $\Gamma_n : \mathsf{Ab} \to \mathsf{Ab}$ is defined, for $A \in \mathsf{Ab}$, to the nth homogeneous component of the graded group $\Gamma_*(A)$ generated by symbols $\gamma_i(x)$ of degree $i \geq 0$ satisfying the following relations for all x, $y \in A$:

(i) $\gamma_0(x) = 1$,

(ii) $\gamma_1(x) = x$,

(iii) $\gamma_s(x)\gamma_t(x) = \binom{s+t}{s}\gamma_{s+t}(x)$,

(iv) $\gamma_n(x+y) = \sum\limits_{s+t=n} \gamma_s(x)\gamma_t(y),\ n \geq 1$,

(v) $\gamma_n(-x) = (-1)^n\gamma_n(x),\ n \geq 1$.

In particular, the canonical map $A \to \Gamma_1(A)$ is an isomorphism. It is known that, for a free abelian group A, there is a natural isomorphism

$$\Gamma_n(A) \simeq (A^{\otimes n})^{\Sigma_n},\ n \geq 1$$

where the action of the symmetric group Σ_n on $A^{\otimes n}$ is defined as follows

$$\sigma(x_1 \otimes \cdots \otimes x_n) = x_{\sigma(1)} \otimes \cdots \otimes x_{\sigma(n)},\ x_i \in A,\ \sigma \in \Sigma_n. \tag{2.1}$$

Observe that, for an abelian group A, there is a natural exact sequence

$$0 \to A \otimes \mathbb{Z}/2 \to \tilde{\otimes}^2(A) \to \Lambda^2(A) \to 0.$$

We will use the same notation for functors on Ab and for their restriction on fAb. To distinguish Hom and Ext groups for functors on Ab and fAb, we will use the notation $\mathrm{Hom}(F,G)$ (resp. $\mathrm{Ext}(F,G)$) for ordinal natural transformations (resp. extensions) of functors $F, G : \mathsf{Ab} \to \mathsf{Ab}$ and $\mathrm{Hom}_f(F,G)$ (resp. $\mathrm{Ext}_f(F,G)$) for functors $F, G : \mathsf{fAb} \to \mathsf{Ab}$.

Let $F : \mathsf{Ab} \to \mathsf{Ab}$ be a functor. Recall that the *cross-effects* of F are multi-functors defined as

$$F(X_1|\cdots|X_n) = \ker\{F(X_1 \oplus \cdots \oplus X_n) \to \oplus_{i=1}^n F(X_1 \oplus \cdots \hat{X}_i \cdots \oplus X_n)\},$$
$$X_i \in \mathsf{Ab},\ n \geq 2 \tag{2.2}$$

where the maps $F(X_1 \oplus \cdots \oplus X_n) \to F(X_1 \oplus \cdots \hat{X}_i \cdots \oplus X_n)$ are induced by natural retractions. The functor F is polynomial of degree d ($d \geq 1$) if $F(0) = 0$ and $F(X_1|\cdots|X_d)$ is linear in each variable $X_i,\ i = 1, \ldots, d$.

Given a functor F and an abelian group A, consider the system of abelian groups:

$$F_1 = F(A),\ F_2 = F(A|A), \ldots, F_n = F(A|\ldots|A) \ (n \text{ copies of } A)$$

together with the homomorphisms

$$H_m^n : F_n(A) \to F_{n+1}(A),\ P_m^n : F_{n+1}(A) \to F_n(A), m = 1, 2, \ldots, m < n$$

which are defined as composite maps

$$H_m^{n+1} : F_n(A) \hookrightarrow F(A^{\oplus n}) \to F(A^{\oplus n+1}) \twoheadrightarrow F_{n+1}(A)$$
$$P_m^{n+1} : F_{n+1}(A) \to F(A^{\oplus n+1}) \to F(A^{\oplus n}) \to F_n(A),$$

induced by natural maps $A^{\oplus n} \to A^{\oplus n+1},\ A^{\oplus n+1} \to A^{\oplus n}$ given by

$$(a_1, \ldots, a_n) \to (a_1, \ldots, a_m, a_m, \ldots, a_n)$$
$$(a_1, \ldots, a_{n+1}) \to (a_1, \ldots, a_{m-1}, a_m + a_{m+1}, a_{m+2}, \ldots, a_{n+1}).$$

Denote the system of these abelian groups and maps by $J_F(A)$:

$$J_F(A) :\ F_1(A) \rightleftarrows F_2(A) \rightleftarrows F_3(A) \rightleftarrows \cdots$$

These maps satisfy certain standard relations [4], which do not depend on F and A. For a polynomial functor F of degree d and an abstract collection of d abelian groups $\{F_i(\mathbb{Z})\}_{i=1,\ldots,d}$ together with corresponding maps which satisfy these relations is known as d-polynomial $\mathbb{Z}$-module. Polynomial functors from free abelian groups to abelian groups can be described in terms of polynomial $\mathbb{Z}$-modules [4]. We now consider some particular cases.

2.1. Quadratic functors

In the case of quadratic functors (see [2]), the required relations are simple:

$$A_1 \xrightarrow{H_1^2} A_2 \xrightarrow{P_1^2} A_1$$
$$H_1^2 P_1^2 H_1^2 = 2H_1^2,\ \ P_1^2 H_1^2 P_1^2 = 2P_1^2. \tag{2.3}$$

Such a diagram of abelian groups is called a quadratic $\mathbb{Z}$-module. It is easy to compute the quadratic $\mathbb{Z}$-modules, which correspond to the classical quadratic functors mentioned above and to $-\otimes \mathbb{Z}/2$. Here they are:

$$
\begin{aligned}
\mathbb{Z}^{\otimes^2} &= (\mathbb{Z} \overset{(1,1)}{\to} \mathbb{Z}\oplus\mathbb{Z} \overset{(1,1)}{\to} \mathbb{Z})\\
\mathbb{Z}^{\Lambda^2} &= (0 \to \mathbb{Z} \to 0)\\
\mathbb{Z}^{\Gamma_2} &= (\mathbb{Z} \overset{1}{\to} \mathbb{Z} \overset{2}{\to} \mathbb{Z})\\
\mathbb{Z}^{S^2} &= (\mathbb{Z} \overset{2}{\to} \mathbb{Z} \overset{1}{\to} \mathbb{Z})\\
\mathbb{Z}^{\tilde{\otimes}^2} &= (\mathbb{Z}/2 \overset{0}{\to} \mathbb{Z} \overset{1}{\to} \mathbb{Z}/2)\\
\mathbb{Z}^{\mathbb{Z}/2} &= (\mathbb{Z}/2 \to 0 \to \mathbb{Z}/2).
\end{aligned}
$$

Given a quadratic $\mathbb{Z}$-module M (2.3), one can define a quadratic functor on the category of abelian groups as follows (see 6.13 [2]): for an abelian group A, $A\otimes M$ is generated by the symbols $a\otimes m, \{a,b\}\otimes n,\ a,b\in A,\ m\in A_1, n\in A_2$ with the relations

$$
\begin{aligned}
&(a+b)\otimes m = a\otimes m + b\otimes m + \{a,b\}\otimes H_1^2(m),\\
&\{a,a\}\otimes n = a\otimes P_1^2(n),\\
&a\otimes m \text{ is linear in } m,\\
&\{a,b\}\otimes n \text{ is linear in } a,b,n.
\end{aligned}
$$

The correspondence $A\mapsto A\otimes M$ defines a quadratic functor and an equivalence between categories of quadratic $\mathbb{Z}$-modules and quadratic functors $\mathsf{fAb}\to\mathsf{Ab}$.

2.2. Cubical functors

The cubical $\mathbb{Z}$-module is given by the diagram

$$
A_1 \underset{\overset{\longleftarrow}{P_1^2}}{\overset{H_1^2}{\longrightarrow}} A_2 \;\begin{matrix} \xrightarrow{H_2^3} \\ \xrightarrow{H_1^3} \\ \xleftarrow{P_2^3} \\ \xleftarrow{P_1^3} \end{matrix}\; A_3
$$

with the following relations (see [4], [15]):

$$
\begin{aligned}
H_1^3H_1^2 = H_2^3H_1^2,\ P_1^2P_1^3 &= P_1^2P_2^3,\ H_2^3P_1^3 = 0,\\
H_1^3P_2^3 = 0,\ H_1^3P_1^3H_1^3 &= 2H_1^3,\ P_1^3H_1^3P_1^3 = 2P_1^3,\\
H_2^3P_2^3H_2^3 &= 2H_2^3,\ P_2^3H_2^3P_2^3 = 2P_2^3,\\
H_1^2P_1^2H_1^2 &= 2H_1^2 + 2(P_1^3+P_2^3)H_1^3H_1^2,\\
P_1^2H_1^2P_1^2 &= 2P_1^2 + 2P_1^2P_1^3(H_2^3+H_1^3),\\
H_1^3H_1^2P_1^2 + H_1^3 + H_2^3 &= H_2^3P_2^3H_1^3P_1^3H_2^3 + H_1^3P_1^3H_2^3P_2^3H_1^3,\\
H_1^2P_1^2P_1^3 + P_1^3 + P_2^3 &= P_2^3H_1^3P_1^3H_2^3P_2^3 + P_1^3H_2^3P_2^3H_1^3P_1^3.
\end{aligned}
$$

The simplest examples of the cubical $\mathbb{Z}$-modules, which correspond to the exterior and symmetric cubes are the following:

$$\Lambda^3 \rightsquigarrow \qquad 0 \rightleftarrows 0 \overset{\rightrightarrows}{\leftleftarrows} \mathbb{Z} \tag{2.4}$$

$$S^3 \rightsquigarrow \qquad \mathbb{Z} \underset{\overleftarrow{(1,1)}}{\overset{(3,3)}{\rightleftarrows}} \mathbb{Z}\oplus\mathbb{Z} \underset{\overleftarrow{(0,1)},\ \overleftarrow{(1,0)}}{\overset{(0,2),\ (2,0)}{\rightrightarrows}} \mathbb{Z}. \tag{2.5}$$

2.3. Δ-properties

For any functor F, the sequence

$$\mathcal{F}(\mathbb{Z}): \quad F_1(\mathbb{Z}) \overset{P_1^2}{\longleftarrow} F_2(\mathbb{Z}) \underset{\overleftarrow{P_1^3}}{\overset{P_2^3}{\leftleftarrows}} F_3(\mathbb{Z}) \overset{\leftarrow}{\underset{\leftarrow}{\cdots}} \cdots$$

is a Δ-group, that is, the standard simplicial relations for the face maps are satisfied. Taking the homology of this complex, we obtain the values of the derived functors in the sense of Dold–Puppe [14]:

$$H_i(\mathcal{F}(\mathbb{Z})) \simeq L_{i+1}F(\mathbb{Z},1).$$

This follows from the fact that the cross-effect spectral sequence from [14] degenerates to the complex $\mathcal{F}(\mathbb{Z})$.

2.4. Natural transformations between functors

All natural transformations between quadratic functors $\mathsf{fAb} \to \mathsf{Ab}$ are given as morphisms of corresponding quadratic $\mathbb{Z}$-modules. One can therefor use quadratic $\mathbb{Z}$-modules for the computation of the group of natural transformations between given quadratic functors.

Examples. 1. A natural map $S^2(A) \to \Gamma_2(A)$ is given by the following diagram[3]:

$$\begin{array}{ccccc} \mathbb{Z} & \xrightarrow{2} & \mathbb{Z} & \xrightarrow{1} & \mathbb{Z} \\ \downarrow{\scriptstyle 2k} & & \downarrow{\scriptstyle k} & & \downarrow{\scriptstyle 2k} \\ \mathbb{Z} & \xrightarrow{1} & \mathbb{Z} & \xrightarrow{2} & \mathbb{Z} \end{array}$$

for $k \in \mathbb{Z}$, and $\mathrm{Hom}_f(S^2, \Gamma_2) = \mathbb{Z}$.

---

[3]For a map between cyclic groups $f : A \to B$, we will use the notation $A \overset{n}{\to} B$ if $f(a) = nb$, where a and b are some given generators of A and B. Analogously we describe the maps between finitely-generated abelian groups by integral matrices.

2. The natural map $\Gamma_2(A) \to A \otimes \mathbb{Z}/2$ is given by the following diagram:

$$\begin{array}{ccccc} \mathbb{Z} & \xrightarrow{1} & \mathbb{Z} & \xrightarrow{2} & \mathbb{Z} \\ \downarrow & & \downarrow & & \downarrow \\ \mathbb{Z}/2 & \longrightarrow & 0 & \longrightarrow & \mathbb{Z}/2\,. \end{array}$$

3. Let us now prove that there do not exist non-zero natural maps

$$\Lambda^2(A) \to S^2(A), \quad S^2(A) \to A \otimes \mathbb{Z}/2. \tag{2.6}$$

Since the functors in (2.6) are right exact, it is enough to consider these functors on the category of free abelian groups. Hence we can look at morphisms between the corresponding quadratic $\mathbb{Z}$-modules. To every map $\Lambda^2 \to S^2$ corresponds a morphism of quadratic $\mathbb{Z}$-modules:

$$\begin{array}{ccccc} 0 & \longrightarrow & \mathbb{Z} & \longrightarrow & 0 \\ \downarrow & & \downarrow & & \downarrow \\ \mathbb{Z} & \xrightarrow{2} & \mathbb{Z} & \xrightarrow{1} & \mathbb{Z}\,. \end{array}$$

We see that the middle vertical map must be zero, hence the result. Same reasoning applies to the natural transformation $S^2(A) \to A \otimes \mathbb{Z}/2$:

$$\begin{array}{ccccc} \mathbb{Z} & \xrightarrow{2} & \mathbb{Z} & \xrightarrow{1} & \mathbb{Z} \\ \downarrow & & \downarrow & & \downarrow \\ \mathbb{Z}/2 & \longrightarrow & 0 & \longrightarrow & \mathbb{Z}/2\,. \end{array}$$

We see that any such vertical map is zero. Hence, there is no any non-zero natural transformations (2.6).

4. Consider the case of cubical functors. The functors S^3 and Λ^3 are right exact, so that in order to prove that $\mathrm{Hom}(S^3, \Lambda^3) = 0$ it is enough to show that $\mathrm{Hom}_f(S^3, \Lambda^3) = 0$, i.e., that any map between cubical $\mathbb{Z}$-modules 2.5 and 2.4 is zero. It is easy to see that all vertical maps in the following commutative diagrams are zero:

$$\begin{array}{ccccc} \mathbb{Z} & \overset{(3,3)}{\underset{(1,1)}{\rightleftarrows}} & \mathbb{Z}\oplus\mathbb{Z} & \begin{smallmatrix}(0,2)\\(2,0)\\(0,1)\\(1,0)\end{smallmatrix} & \mathbb{Z} \\ \downarrow & & \downarrow & & \downarrow \\ 0 & \rightleftarrows & 0 & \rightleftarrows\rightleftarrows & \mathbb{Z} \end{array} \qquad \begin{array}{ccccc} 0 & \rightleftarrows & 0 & \rightleftarrows\rightleftarrows & \mathbb{Z} \\ \downarrow & & \downarrow & & \downarrow \\ \mathbb{Z} & \overset{(3,3)}{\underset{(1,1)}{\rightleftarrows}} & \mathbb{Z}\oplus\mathbb{Z} & \begin{smallmatrix}(0,2)\\(2,0)\\(0,1)\\(1,0)\end{smallmatrix} & \mathbb{Z} \end{array}$$

This case is very simple due to the structure of cross-effects of Λ^3 and can be easily extended to high-dimensional symmetric and exterior powers.

We collect the Hom-functors between main quadratic functors in the category of free abelian groups in the following table:

$G \setminus F$	Γ_2	$\otimes^2$	$\tilde{\otimes}^2$	$\mathbb{Z}/2$	S^2	Λ^2	$\Lambda^2 \otimes \mathbb{Z}/2$
Γ_2	$\mathbb{Z}$	$\mathbb{Z}$	0	0	$\mathbb{Z}$	0	0
$\otimes^2$	$\mathbb{Z}$	$\mathbb{Z}\oplus\mathbb{Z}$	$\mathbb{Z}$	0	$\mathbb{Z}$	$\mathbb{Z}$	0
$\tilde{\otimes}^2$	$\mathbb{Z}/2$	$\mathbb{Z}$	$\mathbb{Z}$	$\mathbb{Z}/2$	0	$\mathbb{Z}$	0
$\mathbb{Z}/2$	$\mathbb{Z}/2$	0	0	$\mathbb{Z}/2$	0	0	0
S^2	$\mathbb{Z}$	$\mathbb{Z}$	0	0	$\mathbb{Z}$	0	0
Λ^2	0	$\mathbb{Z}$	$\mathbb{Z}$	0	0	$\mathbb{Z}$	0
$\Lambda^2 \otimes \mathbb{Z}/2$	0	$\mathbb{Z}/2$	$\mathbb{Z}/2$	0	$\mathbb{Z}/2$	$\mathbb{Z}/2$	$\mathbb{Z}/2$

Table 1. $\mathrm{Hom}_f(F, G)$

Note that some polynomial functors F of degree n have the property that $\mathrm{Hom}(F, G) = \mathrm{Hom}(G, F) = 0$ for any functor G of degree less than n. Let us now consider examples of functors which do not satisfy this property.

1. For an abelian group A, we have a natural map

$$\Gamma_2(A) \to A \otimes \mathbb{Z}/2.$$

The kernel $K(A)$ of the inclusion map

$$\Gamma_2(A) \to A \otimes A$$

defines a functor K in the category of abelian groups (see [5] for the description of this functor). A simple analysis shows that K is a linear functor.

2. There are natural transformations

$$\mathrm{Tor}(A, \mathbb{Z}/2) \to S^2(A), \quad a \mapsto a^2, \ a \in A, \ 2a = 0$$

and

$$A \otimes \mathrm{Tor}(A, \mathbb{Z}/2) \to S^3(A), \quad a \otimes b \mapsto ab^2, \ a, b \in A, \ 2b = 0.$$

Now observe, that for any functor F from the set $\{\Lambda^n, \ \Lambda^n \otimes \mathbb{Z}/p, \ \otimes^n, \ \otimes^n \otimes \mathbb{Z}/p \quad (p \text{ is a prime})\}$, the natural map

$$F(A) \to F(A|\dots|A) \ (n\text{th cross effect})$$

induced by the diagonal embedding $A \hookrightarrow A \oplus \cdots \oplus A$ (n copies of A) is injective. It follows that

$$\mathrm{Hom}(G, F) = 0 \ \text{ for every functor } G \text{ of degree less than } n. \qquad (2.7)$$

Similarly, the natural projection

$$A \oplus \cdots \oplus A \mapsto A, \ (a_1, \dots, a_n) \mapsto a_1 + \cdots + a_n, \ a_i \in A,$$

induces a natural epimorphism

$$F(A|\dots|A) \to F(A)$$

hence $\mathrm{Hom}(F,G) = 0$ for every functor G of degree less than n.

2.5. Extensions between functors

A natural short exact sequence

$$0 \to S^2(A) \otimes \mathbb{Z}/2 \to \Gamma_2(A \otimes \mathbb{Z}/4) \to \Gamma_2(A \otimes \mathbb{Z}/2) \to 0$$

is given by the following short exact sequence of quadratic $\mathbb{Z}$-modules[4]:

$$\begin{array}{ccccc} \mathbb{Z}/2 & \xrightarrow{0} & \mathbb{Z}/2 & \xrightarrow{1} & \mathbb{Z}/2 \\ \downarrow & & \downarrow & & \downarrow \\ \mathbb{Z}/8 & \twoheadrightarrow & \mathbb{Z}/4 & \rightarrowtail & \mathbb{Z}/8 \\ \downarrow & & \downarrow & & \downarrow \\ \mathbb{Z}/4 & \twoheadrightarrow & \mathbb{Z}/2 & \rightarrowtail & \mathbb{Z}/4. \end{array} \tag{2.8}$$

Similarly, a natural exact sequence

$$0 \to \Lambda^2(A) \otimes \mathbb{Z}/2 \to \Gamma_2(A) \otimes \mathbb{Z}/2 \to A \otimes \mathbb{Z}/2 \to 0$$

is given by

$$\begin{array}{ccccc} 0 & \longrightarrow & \mathbb{Z}/2 & \longrightarrow & 0 \\ \downarrow & & \downarrow{\scriptstyle 1} & & \downarrow \\ \mathbb{Z}/2 & \xrightarrow{1} & \mathbb{Z}/2 & \xrightarrow{0} & \mathbb{Z}/2 \\ \downarrow & & \downarrow & & \downarrow \\ \mathbb{Z}/2 & \longrightarrow & 0 & \longrightarrow & \mathbb{Z}/2. \end{array}$$

The following proposition follows directly from the structure of polynomial $\mathbb{Z}$-module which corresponds to the exterior power.

Proposition 2.1. *Let F be a functor of degree d, then*

$$\mathrm{Ext}_f(F, \Lambda^{d+2}) = \mathrm{Ext}_f(\Lambda^{d+2}, F) = 0.$$

[4] we will always display quadratic $\mathbb{Z}$-modules horizontally

Using the language of quadratic $\mathbb{Z}$-modules, one can compute the values of the Ext-functors for the main quadratic functors $F, G : \mathsf{fAb} \to \mathsf{Ab}$. For example,

$$\begin{aligned}
&\mathrm{Ext}_f(\Lambda^2 \otimes \mathbb{Z}/2, \Gamma_2) = \mathbb{Z}/2, \\
&\mathrm{Ext}_f(- \otimes \mathbb{Z}/2, S^2) = \mathbb{Z}/2, \\
&\mathrm{Ext}_f(\Lambda^2 \otimes \mathbb{Z}/2, - \otimes \mathbb{Z}/2) = \mathbb{Z}/2, \\
&\mathrm{Ext}_f(\Gamma_2, S^2) = 0, \\
&\mathrm{Ext}_f(- \otimes \mathbb{Z}/2, \Lambda^2 \otimes \mathbb{Z}/2) = \mathbb{Z}/2, \\
&\mathrm{Ext}_f(\Gamma_2, \Gamma_2) = 0, \\
&\mathrm{Ext}_f(\Gamma_2, \Lambda^2 \otimes \mathbb{Z}/2) = 0\,.
\end{aligned}$$

The proofs are direct, they follow from computations of the extensions between quadratic $\mathbb{Z}$-modules which correspond to the quadratic functors.

The generators of the Ext-groups $\mathrm{Ext}_f(\Lambda^2 \otimes \mathbb{Z}/2, \Gamma_2)$, $\mathrm{Ext}_f(- \otimes \mathbb{Z}/2, S^2)$, $\mathrm{Ext}_f(\Lambda^2 \otimes \mathbb{Z}/2, - \otimes \mathbb{Z}/2)$ one can find in the following diagram

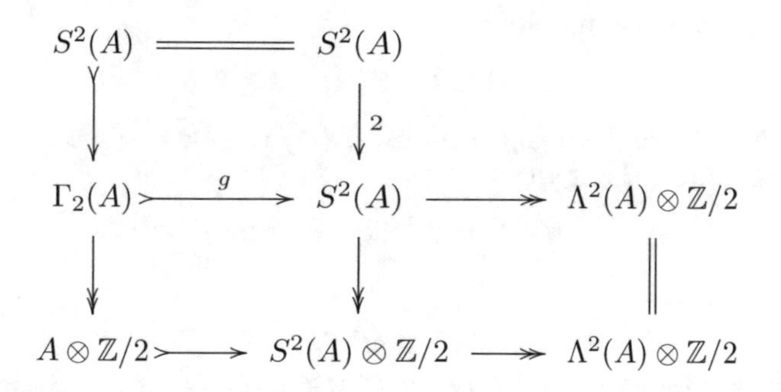

where the map g is given by setting $g : \gamma_2(a) \mapsto a^2,\ a \in A$. It is shown in [8] that

$$\mathrm{Ext}_f(\Gamma_2 \otimes \mathbb{Z}/2, - \otimes \mathbb{Z}/2) = \mathbb{Z}/2 \oplus \mathbb{Z}/2.$$

One of the extensions is given by functor $\Gamma_2(- \otimes/\mathbb{Z}/2)$. This functor together with the generator of $\mathrm{Ext}_f(- \otimes \mathbb{Z}/2, \Lambda^2 \otimes \mathbb{Z}/2)$ one can find in the following diagram:

$$\begin{array}{ccccc}
A \otimes \mathbb{Z}/2 & \rightarrowtail & S^2(A) \otimes \mathbb{Z}/2 & \twoheadrightarrow & \Lambda^2(A) \otimes \mathbb{Z}/2 \\
\| & & \downarrow & & \downarrow \\
A \otimes \mathbb{Z}/2 & \rightarrowtail & \Gamma_2(A \otimes \mathbb{Z}/2) & \twoheadrightarrow & \Gamma_2(A) \otimes \mathbb{Z}/2 \\
& & \downarrow & & \downarrow \\
& & A \otimes \mathbb{Z}/2 & = & A \otimes \mathbb{Z}/2\,.
\end{array}$$

3. Homology of abelian groups

We will now show that (1.2) does not split naturally. For an abelian group A, recall the bar-resolution

$$\mathcal{B}(A): \qquad \dots \to \mathbb{Z}[A \oplus A \oplus A] \xrightarrow{d_3} \mathbb{Z}[A \oplus A] \xrightarrow{d_2} \mathbb{Z}[A] \xrightarrow{d_1} \mathbb{Z}$$

where the differential

$$d_i : \mathbb{Z}[A^{\oplus i}] \to \mathbb{Z}[A^{\oplus i-1}]$$

is given by

$$d_i : (a_1, \dots, a_i) \mapsto (a_1, \dots, a_{i-1}) + \sum_{j=1}^{i-1} (-1)^j (a_1, \dots, a_j + a_{j+1}, \dots, a_i) \\ + (-1)^i (a_2, \dots, a_i).$$

There is a natural isomorphism in the derived category

$$\mathbb{Z}[A[1]] \simeq \mathcal{B}(A)$$

and, in particular, an isomorphism

$$H_i(A) \simeq H_i(\mathcal{B}(A)), \ i \geq 0.$$

Some generators of the homology groups $H_i(A)$ can be easily described in terms of $\mathcal{B}(A)$, for example, the map

$$H_2(A) \simeq \Lambda^2(A) \to \ker(d_2)$$

is given by

$$a \wedge b \mapsto (a, b) - (b, a), \ a, b \in A.$$

Consider the functor $H_2(A; \mathbb{Z}/2) \simeq H_2(\mathcal{B}(A) \otimes \mathbb{Z}/2)$. The universal coefficient theorem implies that there is a natural exact sequence

$$0 \to \Lambda^2(A) \otimes \mathbb{Z}/2 \to H_2(A; \mathbb{Z}/2) \to \mathrm{Tor}(A, \mathbb{Z}/2) \to 0\,. \tag{3.1}$$

Consider the map $f : H_2(A; \mathbb{Z}/2) \to H_2(A|A; \mathbb{Z}/2) = A \otimes A \otimes \mathbb{Z}/2$ induced by the diagonal map $A \to A \oplus A$. Suppose that the sequence (3.1) splits naturally, i.e., $H_2(A; \mathbb{Z}/2) = \Lambda^2(A) \otimes \mathbb{Z}/2 \oplus \mathrm{Tor}(A, \mathbb{Z}/2)$. Then the composition map $\mathrm{Tor}(A, \mathbb{Z}/2) \to H_2(A; \mathbb{Z}/2) \xrightarrow{f} H_2(A|A; \mathbb{Z}/2)$ is zero, since we have seen that there is no non-trivial natural transformation between a linear functor and $A \otimes A \otimes \mathbb{Z}/2$. In particular, for $A = \mathbb{Z}/2$, the map

$$f : H_2(\mathbb{Z}/2; \mathbb{Z}/2) \to H_2(\mathbb{Z}/2|\mathbb{Z}/2; \mathbb{Z}/2)$$

is zero. Let a be a generator of $A = \mathbb{Z}/2$. One has $H_2(A; \mathbb{Z}/2) = \mathbb{Z}/2$ and the generator of this $\mathbb{Z}/2$ in the bar-resolution can be chosen as $(a, a) \in \mathbb{Z}/2[A \oplus A]$. This follows from the fact that $(a, a) \in \ker(d_2) \setminus \mathrm{im}(d_3)$, since $\mathrm{im}(d_3)$ lies in the augmentation ideal of $\mathbb{Z}/2[A \oplus A]$. Taking $B = A = \mathbb{Z}/2$ and b as a generator of B, we now consider the map of bar resolutions $\mathcal{B}(A) \to \mathcal{B}(A \oplus B)$, induced by the diagonal map $A \to A \oplus B$. We see that the image of the map f is generated by an element $(a+b, a+b) \in \mathbb{Z}/2[(A \oplus B) \oplus (A \oplus B)]$. It is easy to verify that

$$(a+b, a+b) + (a, a) + (b, b) \equiv (a, b) + (b, a) \mod \operatorname{im}(d_3)$$

in $\mathbb{Z}/2[(A \oplus B) \oplus (A \oplus B)]$. This implies that the image of the element $(a+b, a+b)$ under the map $H_2(A \oplus B; \mathbb{Z}/2) \to H_2(A|B; \mathbb{Z}/2) = A \otimes B \otimes \mathbb{Z}/2$ is the same as the image of the element $(a, b) + (b, a)$. However, the image of the element $(a, b) + (b, a)$ in $A \otimes B \otimes \mathbb{Z}/2$ is exactly $a \otimes b \otimes 1$, which is the generator of $A \otimes B \otimes \mathbb{Z}/2$. This proves that the sequence (3.1) does not split functorially.

Lemma 3.1. *There is a natural isomorphism*

$$H_2(A; \mathbb{Z}/2) \simeq H_3(A|\mathbb{Z}/2).$$

Proof. We have

$$\mathcal{B}(A \oplus \mathbb{Z}/2) \simeq \mathcal{B}(A) \otimes \mathcal{B}(\mathbb{Z}/2).$$

Since

$$\mathcal{B}(\mathbb{Z}/2) \simeq \mathbb{Z} \oplus \bigoplus_{n \geq 0} \mathbb{Z}/2[2n+1],$$

we have a natural isomorphism

$$\begin{aligned} H_3(A \oplus \mathbb{Z}/2) &\simeq H_3(\mathcal{B}(A) \oplus (\mathcal{B}(A) \otimes \mathbb{Z}/2[1]) \oplus (\mathcal{B}(A) \otimes \mathbb{Z}/2[3])) \\ &\simeq H_3(A) \oplus H_2(\mathcal{B}(A) \otimes \mathbb{Z}/2) \oplus H_3(\mathbb{Z}/2), \end{aligned} \tag{3.2}$$

where the natural maps $H_3(A) \to H_3(A \oplus \mathbb{Z}/2)$ and $H_3(\mathbb{Z}/2) \to H_3(A \oplus \mathbb{Z}/2)$ are splitting monomorphisms on the direct summands in (3.2). It follows that

$$H_3(A|\mathbb{Z}/2) \simeq H_2(\mathcal{B}(A) \otimes \mathbb{Z}/2) \simeq H_2(A; \mathbb{Z}/2).$$

□

Corollary 3.1. *The natural sequence* (1.2)

$$0 \to \Lambda^3(A) \to H_3(A) \to \Omega_2(A) \to 0 \tag{3.3}$$

does not split functorially.

Proof. Suppose that the sequence (3.3) splits naturally, i.e., there is a natural isomorphism

$$H_3(A) \simeq \Lambda^3(A) \oplus \Omega_2(A).$$

This induces the following natural decomposition for the cross-effect functor

$$H_3(A|\mathbb{Z}/2) \simeq \Lambda^2(A) \otimes \mathbb{Z}/2 \oplus \operatorname{Tor}(A, \mathbb{Z}/2).$$

Lemma 3.1 implies that there is a natural isomorphism

$$H_2(A; \mathbb{Z}/2) \simeq \Lambda^2(A) \otimes \mathbb{Z}/2 \oplus \operatorname{Tor}(A, \mathbb{Z}/2),$$

however this contradicts the fact that the sequence (3.1) does not split functorially.

□

Observe that, for any free abelian group, there is a natural isomorphism

$$H_2(A \otimes \mathbb{Z}/2; \mathbb{Z}/2) \simeq \Gamma_2(A) \otimes \mathbb{Z}/2$$

and the functor $H_2(A \otimes \mathbb{Z}/2; \mathbb{Z}/2)$ represents the non-trivial element of

$$\mathrm{Ext}_f(- \otimes \mathbb{Z}/2, \Lambda^2 \otimes \mathbb{Z}/2) = \mathbb{Z}/2.$$

As a consequence of Corollary 3.1, for a free abelian group A, the functor

$$A \mapsto H_3(A \otimes \mathbb{Z}/2)$$

lives in the following short exact sequence

$$0 \to \Lambda^3(A) \otimes \mathbb{Z}/2 \to H_3(A \otimes \mathbb{Z}/2) \to \Gamma_2(A) \otimes \mathbb{Z}/2 \to 0$$

and is represented by the following cubical $\mathbb{Z}$-module:

$$\mathbb{Z}/2 \;\underset{\longleftarrow}{\overset{1}{\underset{0}{\longrightarrow}}}\; \mathbb{Z}/2 \;\begin{smallmatrix} \xrightarrow{1} \\ \xrightarrow[0]{1} \\ \xleftarrow[]{} \\ \xleftarrow{0} \end{smallmatrix}\; \mathbb{Z}/2\,.$$

In particular, this functor represents the non-trivial element in the corresponding Ext-group:

$$H_3(A \otimes \mathbb{Z}/2)(\neq 0) \in \mathrm{Ext}_f(\Gamma_2 \otimes \mathbb{Z}/2, \Lambda^3 \otimes \mathbb{Z}/2) = \mathbb{Z}/2. \tag{3.4}$$

We are now ready to generalize Corollary 3.1 to the case of higher homology functors.

Proposition 3.1. *Let A be an abelian group. For $n \geq 3$, the natural monomorphism induced by Pontryagin product*

$$\Lambda^n(A) \hookrightarrow H_n(A)$$

does not split naturally.

Proof. For $n = 3$ this is Corollary 3.1. Now the result follows by induction on n, observing that there is a natural isomorphism

$$H_n(A|\, \mathbb{Z}) \simeq H_{n-1}(A)$$

which follows from the Künneth formula. Indeed, assuming that the monomorphism $\Lambda^n(A) \hookrightarrow H_n(A)$ splits naturally, we get the natural splitting of the cross-effects

$$\begin{array}{ccc} \Lambda^n(A|\, \mathbb{Z}) & \rightarrowtail & H_n(A|\, \mathbb{Z}) \\ \downarrow \simeq & & \downarrow \simeq \\ \Lambda^{n-1}(A) & \rightarrowtail & H_{n-1}(A) \end{array}$$

but the lower map is not split by induction hypothesis. □

Remark 3.1. *For an odd prime p, the functor*

$$A \mapsto H_3(A \otimes \mathbb{Z}/p), \ A \in \mathsf{fAb}$$

splits as

$$H_3(A \otimes \mathbb{Z}/p) = \Lambda^3(A) \otimes \mathbb{Z}/p \oplus \Gamma_2(A) \otimes \mathbb{Z}/p. \tag{3.5}$$

Proof. First we prove that $\mathrm{Ext}_f(\Gamma_2 \otimes \mathbb{Z}/p, \Lambda^3 \otimes \mathbb{Z}/p) = 0$. Every element of this Ext-group can be presented as a diagram of the form

$$\begin{array}{ccccc}
0 & \rightleftarrows & 0 & \overset{\rightrightarrows}{\leftleftarrows} & \mathbb{Z}/p \\
\downarrow & & \downarrow & & \downarrow^{=} \\
\mathbb{Z}/p & \underset{\longleftarrow}{\overset{1}{\underset{2}{\longrightarrow}}} & \mathbb{Z}/p & \begin{smallmatrix} \xrightarrow{h_1} \\ \xrightarrow{h_2} \\ \xrightarrow{p_1} \\ \xleftarrow{p_2} \\ \longleftarrow \end{smallmatrix} & \mathbb{Z}/p \\
\downarrow & & \downarrow & & \downarrow \\
\mathbb{Z}/p & \underset{\longleftarrow}{\overset{1}{\underset{2}{\longrightarrow}}} & \mathbb{Z}/p & \overset{\rightrightarrows}{\leftleftarrows} & 0
\end{array}$$

It follows immediately that p_1 and p_2 are zero maps. The relations

$$h_1 p_1 h_1 = 2h_1, \ h_2 p_2 h_2 = 2h_2$$

imply that h_1 and h_2 are zero map. Hence

$$\mathrm{Ext}_f(\Gamma_2 \otimes \mathbb{Z}/p, \Lambda^3 \otimes \mathbb{Z}/p) = 0.$$

The splitting (3.5) follows from the fact that, for a free abelian A, the sequence (3.3) has the form

$$0 \to \Lambda^3(A) \otimes \mathbb{Z}/p \to H_3(A \otimes \mathbb{Z}/p) \to \Gamma_2(A) \otimes \mathbb{Z}/p \to 0. \qquad \square$$

4. The splitting of the derived functors

4.1. Derived functors

Let A be an abelian group, and F an endofunctor on the category of abelian groups. Recall that for every $n \geq 0$ the derived functor of F in the sense of Dold–Puppe [14] are defined by

$$L_i F(A, n) = \pi_i(FKP_*[n]), \ i \geq 0$$

where $P_* \to A$ is a projective resolution of A, and K is the Dold–Kan transform, the inverse to the Moore normalization functor

$$N : \mathrm{Simpl}(\mathsf{Ab}) \to \mathrm{Chain}(\mathsf{Ab})$$

from simplicial abelian groups to chain complexes. We denote by $LF(A, n)$ the object in the homotopy category of simplicial abelian groups determined by the simplicial abelian group $FK(P_*[n])$, so that

$$L_i F(A, n) = \pi_i(LF(A, n)).$$

We set $LF(A) := LF(A,0)$ and $L_iF(A) := L_iF(A,0)$ for any $i \geq 0$. For a functor F, LF is a functor from $\mathsf{DAb}_{\leq 0}$ to $\mathsf{DAb}_{\leq 0}$. In the next section, a generic element of $\mathsf{DAb}_{\leq 0}$ is denoted by C.

4.2. The splitting of the derived functors of Γ_2

The natural exact sequence

$$0 \to S^2(A) \to \Gamma_2(A) \to A \otimes \mathbb{Z}/2 \to 0$$

implies that, for an object $C \in \mathsf{DAb}_{\leq 0}$, one has a distinguished triangle

$$LS^2(C) \to L\Gamma_2(C) \to C \overset{L}{\otimes} \mathbb{Z}/2 \to LS^2(C)[1]\,. \tag{4.1}$$

Theorem 4.1. *For any $C \in \mathsf{DAb}_{\leq 0}$, there are natural isomorphisms*

$$\pi_i(L\Gamma_2(C[1])) \simeq \pi_i(LS^2(C[1])) \oplus \pi_i\left(C \overset{L}{\otimes} \mathbb{Z}/2[1]\right) \tag{4.2}$$

for all $i \geq 0$.

The following lemma follows from (Satz 12.1 [14]).

Lemma 4.1. *Let $C \in \mathsf{DAb}_{\leq 0}$ be such that $H_0(C) = 0$, then one has $\pi_1(S^2(C)) = 0$. If $H_i(C) = 0$ for $i \leq m$ $(m \geq 1)$, then*

$$\pi_i(LS^2(C)) = 0, \quad i \leq m+2\,.$$

Lemma 4.2. *For every $C \in \mathsf{DAb}_{\leq 0}$, the suspension homomorphism*

$$\pi_1(LS^2(C)) \to \pi_2(LS^2(C[1]))$$

is the zero map.

Proof. We have the following natural diagram

$$\begin{array}{ccc} \pi_1(LS^2(C)) & \xrightarrow{\simeq} & L_1S^2(H_0(C)) \\ \big\downarrow{\scriptstyle \text{susp}} & & \big\downarrow \\ \pi_2(LS^2(C[1])) & \xrightarrow{\simeq} & \Lambda^2(H_0(C))\,. \end{array} \tag{4.3}$$

The right-hand vertical map is zero by (Corollary 6.6, [14]). Another way to see why this map is trivial is to write the cross-effect spectral sequence for $\pi_*(LS^2(C[1]))$ from [14]. The first page of this spectral sequence implies that there is an exact sequence

$$\begin{aligned} 0 \to L_1\Lambda^2(H_0(C)) \to \operatorname{Tor}(H_0(C), H_0(C)) \to L_1S^2(H_0(C)) \\ \to \Lambda^2(H_0(C)) \to H_0(C) \otimes H_0(C) \to S^2(H_0(C)) \to 0 \end{aligned}$$

where the middle map is the map from (4.3). It is the zero map since the natural transformation $\Lambda^2(H_0(C)) \to H_0(C) \otimes H_0(C)$ is injective. $\square$

Proof of Theorem 4.1. The proof is by induction on i. Lemma 4.1 implies that there is a natural isomorphism

$$\pi_1(L\Gamma_2(C[1])) \simeq \pi_1\left(C \overset{L}{\otimes} \mathbb{Z}/2[1]\right)$$

which is induced by the map $L\Gamma_2(C[1]) \to C \overset{L}{\otimes} \mathbb{Z}/2[1]$ from (4.1).

Let us consider separately the case $i = 2$. The assertion follows from the suspension diagram

$$\begin{array}{ccccccc}
\pi_3\left(C\overset{L}{\otimes}\mathbb{Z}/2[1]\right) & \longrightarrow & \pi_2(LS^2(C[1])) & \longrightarrow & \pi_2(L\Gamma_2(C[1])) & \twoheadrightarrow & \pi_2\left(C\overset{L}{\otimes}\mathbb{Z}/2[1]\right) \\
\| & & \uparrow{\scriptstyle 0} & & \uparrow & & \| \\
\pi_2\left(C\overset{L}{\otimes}\mathbb{Z}/2\right) & \longrightarrow & \pi_1(LS^2(C)) & \longrightarrow & \pi_1(L\Gamma_2(C)) & \twoheadrightarrow & \pi_1\left(C\overset{L}{\otimes}\mathbb{Z}/p\right)
\end{array}$$

where the left-hand vertical homomorphism is zero by Lemma 4.2.

Now assume by induction, that, for some $j \geq 2$ and for all $i \geq j$, there are natural isomorphisms (4.2), induced by (4.1). Representing the object C as

$$\cdots \to C_i \overset{\partial_i}{\to} C_{i+1} \to \cdots,$$

consider the subcomplex Z defined by

$$\begin{aligned}
Z_i &= C_i, \ i \geq j-1, \\
Z_{j-2} &= \operatorname{im}(\partial_{i-1}), \\
Z_i &= 0, \ i < j-2\,.
\end{aligned}$$

The complex Z has the following properties:

1) the natural map $Z \to C$ induces isomorphisms

$$\pi_i\left(Z \overset{L}{\otimes} \mathbb{Z}/2\right) \simeq \pi_i\left(C \overset{L}{\otimes} \mathbb{Z}/2\right), \ i \geq j;$$

2) $H_i(Z) = 0, \ i \leq j-2$.

Consider the natural diagram

$$\begin{array}{ccccccc}
\pi_{j+2}\left(C\overset{L}{\otimes}\mathbb{Z}/2[1]\right) & \longrightarrow & \pi_{j+1}(LS^2(C[1])) & \longrightarrow & \pi_{j+1}(L\Gamma_2(C[1])) & \twoheadrightarrow & \pi_{j+1}\left(C\overset{L}{\otimes}\mathbb{Z}/2[1]\right) \\
\| & & \uparrow & & \uparrow & & \| \\
\pi_{j+2}\left(Z\overset{L}{\otimes}\mathbb{Z}/2[1]\right) & \longrightarrow & \pi_{j+1}(LS^2(Z[1])) & \longrightarrow & \pi_{j+1}(L\Gamma_2(Z[1])) & \twoheadrightarrow & \pi_{j+1}\left(Z\overset{L}{\otimes}\mathbb{Z}/2[1]\right).
\end{array} \tag{4.4}$$

Lemma 4.1 implies that $\pi_{j+1}(LS^2(Z[1])) = 0$. The required splitting now follows from diagram (4.4). The inductive step is complete so that the splitting (4.2) is proved for all i. □

Proposition 4.1. *The sequence*

$$LS^2(C[1]) \to L\Gamma_2(C[1]) \to C \overset{L}{\otimes} \mathbb{Z}/2[1] \tag{4.5}$$

does not split in the category $\mathsf{DAb}_{\leq 0}$.

Proof. We will prove the statement for the simplest case, when C is a free abelian group. Suppose that $L\Gamma_2(C[1]) \simeq LS^2(C[1]) \oplus C \overset{L}{\otimes} \mathbb{Z}/2[1]$. Then

$$\begin{aligned}\pi_2\left(L\Gamma_2(C[1]) \overset{L}{\otimes} \mathbb{Z}/2\right) &\simeq \pi_2\left(LS^2(C[1]) \overset{L}{\otimes} \mathbb{Z}/2 \oplus C \overset{L}{\otimes} \mathbb{Z}/2 \overset{L}{\otimes} \mathbb{Z}/2[1]\right)\\ &\simeq \Lambda^2(C) \otimes \mathbb{Z}/2 \oplus C \otimes \mathbb{Z}/2\,.\end{aligned}$$

However, $L\Gamma_2(C[1])$ can be represented by complex

$$(C \otimes C \otimes \mathbb{Z}/2 \to \Gamma_2(C) \otimes \mathbb{Z}/2)[1]$$

with the obvious map, and

$$\pi_2\left(L\Gamma_2(C[1]) \overset{L}{\otimes} \mathbb{Z}/2\right) = \ker\{C \otimes C \otimes \mathbb{Z}/2 \to \Gamma_2(C) \otimes \mathbb{Z}/2\}.$$

However, there is no non-trivial natural transformation $C \otimes \mathbb{Z}/2 \to C \otimes C \otimes \mathbb{Z}/2$. This contradicts the splitting of (4.5). □

5. Stable homotopy groups of $K(A,1)$

5.1. Whitehead's exact sequence

Let X be a $(r-1)$-connected CW-complex, $r \geq 2$. Consider the following long exact sequence of abelian groups [28]:

$$\cdots H_{n+1}X \to \Gamma_n X \to \pi_n(X) \overset{h_n}{\to} H_n X \to \Gamma_{n-1}X \to \cdots, \tag{5.1}$$

where $\Gamma_n X = \mathrm{im}\{\pi_n(X^{n-1}) \to \pi_n(X^n)\}$ (here X^i is the ith skeleton of X), h_n is the nth Hurewicz homomorphism. The Hurewicz theorem is equivalent to the statement $\Gamma_i X = 0,\ i \leq r$. J.H.C. Whitehead computed the term $\Gamma_{r+1}X$ (see [28])[5]:

$$\Gamma_{r+1}X = \begin{cases}\Gamma_2(\pi_2(X)),\ r=2\\ \pi_r(X)\otimes\mathbb{Z}/2,\ r>2\end{cases}$$

where $\Gamma_2 : \mathsf{Ab} \to \mathsf{Ab}$ is the universal quadratic functor (or equivalently the divided square).

Consider the stable analog of the Whitehead exact sequence in low degrees. Here we recall the description of functors $\Gamma_i,\ i = r+1, r+2, r+3$ from [3]. Assume that X is $(r-1)$-connected complex, $r \geq 6$. In this case, we have the following:

$$\eta_1 : \pi_r(X) \otimes \mathbb{Z}/2 \to \pi_{r+1}(X)$$

[5] Care should be taken to distinguish between Whitehead's functors $\Gamma_n X$ and the divided power functors $\Gamma_i(A)$.

is induced by the Hopf map $\eta_r \in \pi_{r+1}(S^r)$, i.e., $\eta^1(\alpha \otimes 1) = \alpha\eta_r$ and there is a natural exact sequence

$$0 \to \pi_{r+1}(X) \otimes \mathbb{Z}/2 \to \Gamma_{r+2}X \to \mathrm{Tor}(\pi_r(X), \mathbb{Z}/2) \to 0$$

where the composite map $\pi_{r+1}(X) \otimes \mathbb{Z}/2 \to \pi_{r+2}(X)$ is induced by the Hopf map $\eta_{r+1} \in \pi_{r+1}(S^r)$.

The description of the term $\Gamma_{r+3}X$ is given as follows. There is a natural exact sequence

$$L_2\Gamma^2(\eta_1) \to \Gamma^3(\eta_1, \eta_2) \to \Gamma_{r+3}X \to L_1\Gamma^2(\eta^1) \to 0, \tag{5.2}$$

where the functors in this sequence can be described in the following way:

$$L_1\Gamma^2(\eta_1) = \mathrm{coker}\{\pi_r(X) \otimes \mathbb{Z}/2 \xrightarrow{\eta_1} \mathrm{Tor}(\pi_{r+1}(X), \mathbb{Z}/2)\}$$
$$L_2\Gamma^2(\eta_1) = \ker(\eta_1)$$

and $\Gamma^3(\eta_1, \eta_2) = \pi_r(X) \otimes \mathbb{Z}/3 \oplus P$, where P is given by the pushout

$$\begin{array}{ccc} \pi_r(X) \otimes \mathbb{Z}/2 & \longrightarrow & \pi_{r+2} \otimes \mathbb{Z}/2 \\ \Big\downarrow {\scriptstyle \pi_r(X)\otimes 4} & & \Big\downarrow \\ \pi_r(X) \otimes \mathbb{Z}/8 & \longrightarrow & P \end{array} \tag{5.3}$$

where the upper horizontal map induced by the map $S^{r+2} \to S^r$, which defines a generator of $\pi_2^S = \mathbb{Z}/2$.

5.2. A spectral sequence

Recall the spectral sequence from [24]. Consider an abelian group A and its two-step flat resolution

$$0 \to A_1 \to A_0 \to A \to 0.$$

By Dold–Kan correspondence, one obtains the following free abelian simplicial resolution of A:

$$N^{-1}(A_1 \hookrightarrow A_0) : \quad \cdots \; \substack{\longrightarrow\\[-1em]\longrightarrow\\[-1em]\longrightarrow\\[-1em]\longleftarrow\\[-1em]\longleftarrow} \; A_1 \oplus s_0(A_0) \; \substack{\longrightarrow\\[-1em]\longrightarrow\\[-1em]\longleftarrow} \; A_0.$$

Applying the Carlsson construction (see [10] or [24] for the detailed description of this construction) to the resolution $N^{-1}(A_1 \hookrightarrow A_0)$, we obtain the following bisimplicial group:

$$\begin{array}{ccccc} F^{N^{-1}(A_1\hookrightarrow A_0)_2}(S^n)_3 & \substack{\to\\[-1em]\to\\[-1em]\to\\[-1em]\to\\[-1em]\leftarrow\\[-1em]\leftarrow\\[-1em]\leftarrow} & F^{N^{-1}(A_1\hookrightarrow A_0)_2}(S^n)_2 & \substack{\to\\[-1em]\to\\[-1em]\to\\[-1em]\leftarrow\\[-1em]\leftarrow} & N^{-1}(A_1 \hookrightarrow A_0)_2 \\ \downarrow\downarrow\downarrow\uparrow\uparrow & & \downarrow\downarrow\downarrow\uparrow\uparrow & & \downarrow\downarrow\downarrow\uparrow\uparrow \\ F^{A_1\oplus s_0(A_0)}(S^n)_3 & \substack{\to\\[-1em]\to\\[-1em]\to\\[-1em]\to\\[-1em]\leftarrow\\[-1em]\leftarrow\\[-1em]\leftarrow} & F^{A_1\oplus s_0(A_0)}(S^n)_2 & \substack{\to\\[-1em]\to\\[-1em]\to\\[-1em]\leftarrow\\[-1em]\leftarrow} & A_1 \oplus s_0(A_0) \\ \downarrow\downarrow\uparrow & & \downarrow\downarrow\uparrow & & \downarrow\downarrow\uparrow \\ F^{A_0}(S^n)_3 & \substack{\to\\[-1em]\to\\[-1em]\to\\[-1em]\to\\[-1em]\leftarrow\\[-1em]\leftarrow\\[-1em]\leftarrow} & F^{A_0}(S^n)_2 & \substack{\to\\[-1em]\to\\[-1em]\to\\[-1em]\leftarrow\\[-1em]\leftarrow} & A_0 \end{array}$$

Here the mth horizontal simplicial group is the Carlsson construction

$$F^{N^{-1}(A_1 \hookrightarrow A_0)_m}(S^n).$$

By the result of Quillen [25], we obtain the following spectral sequence:

$$E^2_{p,q} = \pi_q(\pi_p \Sigma^n K(N^{-1}(A_1 \hookrightarrow A_0), 1)) \Longrightarrow \pi_{p+q}\Sigma^n K(A, 1). \tag{5.4}$$

In particular, for n sufficiently large, the spectral sequence becomes

$$E^2_{p,q} = \pi_q(\pi_p^S K(N^{-1}(A_1 \hookrightarrow A_0), 1)) \Longrightarrow \pi^S_{p+q} K(A, 1), \tag{5.5}$$

where π_n^S is the nth stable homotopy group.

5.3. The third stable homotopy group of $K(A, 1)$

We now apply the above results for the description of the stable homotopy groups of $K(A, 1)$ in low degrees. Given an abelian group A, the homotopy functors $\pi_*^S : \mathsf{Ab} \to \mathsf{Ab},\ A \mapsto \pi_*^S K(A, 1)$ can be viewed as parts of the Whitehead exact sequence, which functorially depends on A. Recall that, for $r \geq 2$, $\pi_2^S K(A, 1) = \pi_{r+2}\Sigma^r K(A, 1)$ is the antisymmetric square, and the Whitehead sequence has the form [9]:

$$\begin{array}{ccccc}
\Gamma_{r+2}\Sigma^r K(A,1) & \rightarrowtail & \pi_{r+2}\Sigma^r K(A,1) & \twoheadrightarrow & H_2 K(A,1) \\
\| & & \| & & \| \\
A \otimes \mathbb{Z}/2 & \rightarrowtail & A \tilde{\otimes} A & \twoheadrightarrow & \Lambda^2(A).
\end{array}$$

Now consider the next step, the functor $\pi_3^S K(A, 1) = \pi_{r+3}\Sigma^r K(A, 1)$ for $r \geq 4$. First consider the case of a free abelian group A. Observe that, for a free abelian A, one has a natural isomorphism

$$A \tilde{\otimes} A \otimes \mathbb{Z}/2 \simeq S^2(A) \otimes \mathbb{Z}/2.$$

We have the following exact sequence

$$\begin{array}{ccccccc}
H_4(A) & \longrightarrow & \Gamma_{r+3}\Sigma^r K(A,1) & \longrightarrow & \pi_3^S K(A,1) & \twoheadrightarrow & H_3(A) \\
\| & & \| & & \| & & \| \\
\Lambda^4(A) & \longrightarrow & S^2(A) \otimes \mathbb{Z}/2 & \longrightarrow & \pi_3^S K(A,1) & \twoheadrightarrow & \Lambda^3(A).
\end{array}$$

Now observe that any natural transformation $\Lambda^4(A) \to S^2(A) \otimes \mathbb{Z}/2$ is zero, since, for all $n \geq 2$, there is no non-trivial transformations between $\Lambda^n(A)$ and any functor of degree less than n. Therefore, the functor $\pi_3^S : \mathsf{fAb} \to \mathsf{Ab}$ lives in the following exact sequence

$$0 \to S^2(A) \otimes \mathbb{Z}/2 \to \pi_3^S K(A, 1) \to \Lambda^3(A) \to 0. \tag{5.6}$$

It follows from a simple analysis of the extensions of the cubical $\mathbb{Z}$-modules which correspond to the functors $S^2 \otimes \mathbb{Z}/2$ and Λ^3 that any nontrivial extension between these functors can be given by a diagram of the form

$$\begin{array}{ccccc}
\mathbb{Z}/2 & \underset{\longleftarrow}{\overset{0}{\underset{1}{\rightarrow}}} & \mathbb{Z}/2 & \rightleftarrows & 0 \\
\downarrow^{=} & & \downarrow^{=} & & \downarrow \\
\mathbb{Z}/2 & \underset{\longleftarrow}{\overset{0}{\underset{1}{\rightarrow}}} & \mathbb{Z}/2 & \begin{array}{c} \overset{0}{\rightarrow} \\ \overset{0}{\rightarrow} \\ \overset{1}{\rightarrow} \\ \overset{1}{\leftarrow} \\ \leftarrow \end{array} & \mathbb{Z} \\
\downarrow & & \downarrow & & \downarrow^{=} \\
0 & \rightleftarrows & 0 & \rightleftarrows & \mathbb{Z}
\end{array}$$

and

$$\mathrm{Ext}_f(\Lambda^3, S^2 \otimes \mathbb{Z}/2) = \mathbb{Z}/2. \tag{5.7}$$

We will show now that the extension (5.6) presents a non-trivial element of (5.7).

Theorem 5.1. *The functor*

$$\pi_3^S : \mathsf{fAb} \to \mathsf{Ab}, \ A \mapsto \pi_3^S K(A, 1)$$

is given by the following cubical module:

$$\mathbb{Z}/2 \underset{\longleftarrow}{\overset{0}{\underset{1}{\rightarrow}}} \mathbb{Z}/2 \begin{array}{c} \overset{0}{\rightarrow} \\ \overset{0}{\rightarrow} \\ \overset{1}{\rightarrow} \\ \overset{1}{\leftarrow} \\ \leftarrow \end{array} \mathbb{Z}.$$

Proof. Assume that, for a free abelian group A, the functor $\pi_3^S K(A, 1)$ presents the zero element in (5.7), i.e., $\pi_3^S K(A, 1) = S^2(A) \otimes \mathbb{Z}/2 \oplus \Lambda^3(A)$, and let B be a non-free abelian group. The spectral sequence (5.5) implies that there is a natural exact sequence

$$0 \to S^2(B) \otimes \mathbb{Z}/2 \oplus \Lambda^3(B) \to \pi_3^S K(B, 1) \to L_1\tilde{\otimes}^2(B) \to 0\,. \tag{5.8}$$

Consider now the functor

$$\pi_3^S K(- \otimes \mathbb{Z}/2, 1) : \ \mathsf{fAb} \to \mathsf{Ab}, \quad A \mapsto \pi_3^S K(A \otimes \mathbb{Z}/2, 1)\,.$$

There is the following short exact sequence (see [24]):

$$0 \to A \otimes \mathbb{Z}/2 \to L_1\tilde{\otimes}^2(A \otimes \mathbb{Z}/2) \to \Gamma_2(A) \otimes \mathbb{Z}/2 \to 0$$

and $L_1\tilde{\otimes}^2(\mathbb{Z}/2) = \mathbb{Z}/4$. Hence $L_1\tilde{\otimes}^2(A \otimes \mathbb{Z}/2)$ describes a nontrivial element of

$$\mathrm{Ext}(\Gamma_2(A) \otimes \mathbb{Z}/2, A \otimes \mathbb{Z}/2) = \mathbb{Z}/2 \oplus \mathbb{Z}/2,$$

see [8]. One can check that

$$L_1\tilde{\otimes}^2(A \otimes \mathbb{Z}/2) \simeq \Gamma_2(A \otimes \mathbb{Z}/2)$$

and this functor is represented by the quadratic module

$$\mathbb{Z}/4 \twoheadrightarrow Z/2 \hookrightarrow \mathbb{Z}/4.$$

This follows from the presentation of the functor $\tilde{\otimes}^2$ as a quotient $S^2 \to \otimes^2 \twoheadrightarrow \tilde{\otimes}^2$ and the corresponding exact sequence for the derived functors.

Therefore, the sequence (5.8) can be rewritten for $B = A \otimes \mathbb{Z}/2$ as

$$0 \to S^2(A) \otimes \mathbb{Z}/2 \oplus \Lambda^3(A) \otimes \mathbb{Z}/2 \to \pi_3^S K(A \otimes \mathbb{Z}/2, 1) \to \Gamma_2(A \otimes \mathbb{Z}/2) \to 0\,. \quad (5.9)$$

We know from [21] that $\pi_3^S K(\mathbb{Z}/2, 1) = \mathbb{Z}/8$. The diagram of cubical $\mathbb{Z}$-modules which correspond to the extension (5.9) has the following form

$$\begin{array}{ccccc}
\mathbb{Z}/2 & \substack{\xrightarrow[1]{0} \\ \longleftarrow} & \mathbb{Z}/2 & \substack{\xrightarrow{0} \\ \xrightarrow{0} \\ \xrightarrow{0} \\ \xleftarrow{0} \\ \longleftarrow} & \mathbb{Z}/2 \\
\downarrow & & \downarrow & & \downarrow{=} \\
\mathbb{Z}/8 & \substack{\xrightarrow[2]{1} \\ \longleftarrow} & \mathbb{Z}/4 & \substack{\xrightarrow{0} \\ \xrightarrow{0} \\ \xrightarrow{0} \\ \xleftarrow{0} \\ \longleftarrow} & \mathbb{Z}/2 \\
\downarrow & & \downarrow & & \downarrow \\
\mathbb{Z}/4 & \substack{\xrightarrow[2]{1} \\ \longleftarrow} & \mathbb{Z}/2 & \rightrightarrows & 0\,.
\end{array}$$

One verifies that the above extension is unique and therefore,

$$\pi_3^S(A \otimes \mathbb{Z}/2, 1) \simeq \Gamma_2(A \otimes \mathbb{Z}/4) \oplus \Lambda^3(A) \otimes \mathbb{Z}/2$$

(see (2.8)). Observe also that the Whitehead sequence implies that the Hurewicz map

$$\pi_3^S K(A \otimes \mathbb{Z}/2, 1) \to H_3(A \otimes \mathbb{Z}/2)$$

is a natural surjection, which induces isomorphism on the triple cross-effects. However, it is not possible to construct a commutative diagram of the form

$$\begin{array}{ccccc}
\mathbb{Z}/8 & \substack{\xrightarrow[2]{1} \\ \longleftarrow} & \mathbb{Z}/4 & \substack{\xrightarrow{0} \\ \xrightarrow{0} \\ \xrightarrow{0} \\ \xleftarrow{0} \\ \longleftarrow} & \mathbb{Z}/2 \\
\downarrow & & \downarrow & & \downarrow{\simeq} \\
\mathbb{Z}/2 & \substack{\xrightarrow[0]{1} \\ \longleftarrow} & \mathbb{Z}/2 & \substack{\xrightarrow{1} \\ \xrightarrow{1} \\ \xrightarrow{0} \\ \xleftarrow{0} \\ \longleftarrow} & \mathbb{Z}/2\,.
\end{array}$$

This gives a contradiction. Therefore, the functor $\pi_3^S K(A, 1)$ describes a non-trivial element of (5.7). $\square$

Theorem 5.1 implies that the functor $\pi_3^S K(-\mathbb{Z}/2, 1) : \mathsf{fAb} \to \mathsf{Ab}$ is represented by the cubical $\mathbb{Z}$-module

$$\mathbb{Z}/8 \substack{\xrightarrow[2]{1} \\ \longleftarrow} \mathbb{Z}/4 \substack{\xrightarrow{1} \\ \xrightarrow{1} \\ \xrightarrow{2} \\ \xleftarrow{2} \\ \longleftarrow} \mathbb{Z}/2\,.$$

The portion of the Whitehead sequence which contains the natural transformation $\pi_3^S \to H_3$ has the following form

$$\begin{array}{ccccc} S^2(A)\otimes\mathbb{Z}/2 & \rightarrowtail & F(A) & \twoheadrightarrow & \Lambda^3(A)\otimes\mathbb{Z}/2 \\ \downarrow & & \downarrow & & \downarrow \\ \Gamma_2(A\otimes\mathbb{Z}/2) & \rightarrowtail & \pi_3^S K(A\otimes\mathbb{Z}/2,1) & \twoheadrightarrow & H_3(A\otimes\mathbb{Z}/2) \\ \downarrow & & \downarrow & & \downarrow \\ A\otimes\mathbb{Z}/2 & \rightarrowtail & \Gamma_2(A\otimes\mathbb{Z}/2) & \twoheadrightarrow & \Gamma_2(A)\otimes\mathbb{Z}/2 \end{array} \tag{5.10}$$

where the functor $F : \mathsf{fAb} \to \mathsf{Ab}$ is given by the cubical $\mathbb{Z}$-module

$$\mathbb{Z}/2 \underset{\longleftarrow}{\overset{0}{\underset{1}{\rightarrow}}} \mathbb{Z}/2 \underset{\overset{\longleftarrow}{\underset{1}{\longleftarrow}}}{\overset{\overset{0}{\rightarrow}}{\overset{0}{\underset{1}{\rightarrow}}}} \mathbb{Z}/2\,.$$

The spectral sequence (5.5) therefore implies the following

Proposition 5.1. *For an abelian group A, there is a natural exact sequence*

$$0 \to L_0F(A) \to \pi_3^S K(A,1) \to L_1\tilde{\otimes}^2(A) \to 0$$

which does not split.

5.4. The fourth stable homotopy group of $K(A,1)$

Consider first the case of a free abelian group A. We have the following diagram

$$\begin{array}{ccccccc} H_5(A) & \longrightarrow & \Gamma_{r+4}\Sigma^r K(A,1) & \longrightarrow & \pi_4^S K(A,1) & \twoheadrightarrow & H_4(A) \\ \| & & \| & & \| & & \| \\ \Lambda^5(A) & \longrightarrow & A\otimes\mathbb{Z}/3\oplus P & \longrightarrow & \pi_4^S K(A,1) & \twoheadrightarrow & \Lambda^4(A) \end{array}$$

where the term P was defined in (5.3). The functor $A\otimes\mathbb{Z}/3\oplus P$ is cubical. The natural transformation $\Lambda^5(A) \to A\otimes\mathbb{Z}/3\oplus P$ is zero and we have a natural short exact sequence

$$0 \to A\otimes\mathbb{Z}/3\oplus P \to \pi_4^S K(A,1) \to \Lambda^4(A) \to 0. \tag{5.11}$$

Here, by (5.3), the functor P is given by the pushout diagram

$$\begin{array}{ccc} A\otimes\mathbb{Z}/2 & \longrightarrow & \pi_3^S K(A,1) \\ \downarrow{\scriptstyle A\otimes 4} & & \downarrow \\ A\otimes\mathbb{Z}/8 & \longrightarrow & P\,. \end{array}$$

It follows that the functor P can be descibed by the cubical $\mathbb{Z}$-module:

$$\mathbb{Z}/8 \;\substack{\xrightarrow{0}\\ \xleftarrow{4}}\; \mathbb{Z}/2 \;\substack{\xrightarrow{0}\\ \xrightarrow{0}\\ \xrightarrow{1}\\ \xleftarrow{1}\\ \longleftarrow}\; \mathbb{Z}/2\,.$$

Theorem 5.2. *The functor*

$$\pi_4^S : \mathsf{fAb} \to \mathsf{Ab},\ A \mapsto \pi_4^S K(A,1)$$

is described by the following quartic $\mathbb{Z}$-module

$$\mathbb{Z}/24 \;\substack{\xrightarrow{0}\\ \xleftarrow{12}}\; \mathbb{Z}/2 \;\substack{\xrightarrow{0}\\ \xrightarrow{0}\\ \xrightarrow{1}\\ \xleftarrow{1}\\ \longleftarrow}\; \mathbb{Z}/2 \;\substack{\xrightarrow{0}\\ \xrightarrow{0}\\ \xrightarrow{0}\\ \xrightarrow{1}\\ \xleftarrow{1}\\ \xleftarrow{1}\\ \longleftarrow}\; \mathbb{Z}\,. \tag{5.12}$$

Proof. The proof is similar to that of Theorem 5.1. Since the quartic $\mathbb{Z}$-module which corresponds to the functor Λ^4 has a simple form, it is easy to see that

$$\mathrm{Ext}_f(\Lambda^4, -\otimes\mathbb{Z}/3\oplus P) = \mathbb{Z}/2 \tag{5.13}$$

and a nontrivial element in (5.13) is given by the quartic $\mathbb{Z}$-module (5.12). It remains to show that the sequence (5.11) does not split naturally.

Let us assume that the sequence (5.11) does split. Recall that

$$\mathrm{Hom}(\Lambda^4\otimes\mathbb{Z}/2, G) = \mathrm{Hom}(G, \Lambda^4\otimes\mathbb{Z}/2) = 0$$

for every cubical functor G. The spectral sequence (5.5) implies that the functor

$$\pi_4^S(-\otimes\mathbb{Z}/2,1):\ \mathsf{fAb}\to\mathsf{Ab},\quad A\mapsto \pi_4^S(A\otimes\mathbb{Z}/2,1)$$

can be represented as a direct sum

$$\pi_4^S(A\otimes\mathbb{Z}/2,1)\simeq Q\oplus\Lambda^4(A)\otimes\mathbb{Z}/2$$

for some cubical functor $Q:\mathsf{fAb}\to\mathsf{Ab}$. The description (5.2) of $\Gamma_{r+4}\Sigma^4K(A\otimes\mathbb{Z}/2,1)$ implies that it is a cubical functor for all r. It follows that the image of the Hurewicz map

$$\pi_4^S K(A\otimes\mathbb{Z}/2,1)\to H_4(A\otimes\mathbb{Z}/2)$$

also has the form $\bar{Q}\oplus\Lambda^4(A)\otimes\mathbb{Z}/2$ for some cubical functor $\bar{Q}$. The diagram (5.10) implies that the Hurewicz map is an epimorphism, hence we obtain that there is a natural isomorphism

$$H_4(A\otimes\mathbb{Z}/2)\simeq\bar{Q}\oplus\Lambda^4(A)\otimes\mathbb{Z}/2.$$

Since we know that there exists a natural exact sequence (see [6])

$$0 \to \Lambda^4(A) \otimes \mathbb{Z}/2 \to H_4(A \otimes \mathbb{Z}/2) \to L_1\Lambda^3(A \otimes \mathbb{Z}/2) \to 0, \tag{5.14}$$

we have an isomorphism $\bar{Q} \simeq L_1\Lambda^3(A \otimes \mathbb{Z}/2)$ and the sequence (5.14) splits. To prove this rigorously, one considers a quartic $\mathbb{Z}$-module associated to $\bar{Q} \oplus \Lambda^4(A) \otimes \mathbb{Z}/2$ and compares it with any possible extension of type (5.14). It remains to observe that

$$H_4(A \otimes \mathbb{Z}/2 | \mathbb{Z}) \simeq H_3(A \otimes \mathbb{Z}/2)$$

so that splitting of (5.14) implies the splitting of

$$0 \to \Lambda^3(A) \otimes \mathbb{Z}/2 \to H_3(A \otimes \mathbb{Z}/2) \to \Gamma_2(A) \otimes \mathbb{Z}/2 \to 0.$$

However, by (3.4), this is not possible. □

6. Functorial spectral sequences

In general, the combinatorics of spectral sequences is very difficult. In many important cases, nobody knows how to compute the differentials of a spectral sequence or how to solve the extension problem. However, in the case when we study homotopy types which depend functorialy on an abelian group A, one can use the power of functoriality to solve the above problems. For the examples in this section, the filtration of a complex by functors gives rise to a *functorial spectral sequences*, in the following sense:

1. All pages of the spectral sequence are filled by functors instead of abstract abelian groups;
2. All differentials of the spectral sequence are natural transformations between functors;
3. the E^∞-page can be glued to the abutment by a sequence of functorial operations.

6.1. Homology of abelian groups

For an abelian group A, consider a free abelian simplicial resolution of $K(A,1)$: $P_* \to K(A,1)$. The filtration of $\mathbb{Z}[P_*]$ by powers of the augmentation ideal determines a spectral sequence

$$E^2_{pq} = L_{p+q}S^q(A,1) \Rightarrow H_{p+q}(A).$$

This spectral sequence is studied in [6]. In particular, it is shown in [6] that this spectral sequence stabilizes at the level of the second page, i.e., all its differentials are trivial. Using décalage, the spectral sequence can be rewritten as

$$E^2_{pq} = L_p\Lambda^q(A) \Rightarrow H_{p+q}(A). \tag{6.1}$$

For any abelian group A and $n \geq 1$, we set

$$\Omega_n(A) := L_{n-1}\Lambda^n(A).$$

The symmetric group Σ_n acts on $\mathrm{Tor}^{[n]}(A) := L_{n-1}\otimes^n(A)$ and the functor Ω_n can be identified with Σ_n^ϵ-invariants in $\mathrm{Tor}^{[n]}(A)$ [6], [18, Theorem 2.3.3]:

$$\Omega_n(A) \simeq (\mathrm{Tor}^{[n]}(A))^{\Sigma_n^\epsilon}.$$

There exists a sequence of homomorphisms (see [7] and [18]):

$$\lambda_h^n : \Gamma_n({}_hA) \to \Omega_n(A) \tag{6.2}$$

for $h \geq 1$ such that the group $\Omega_n(A)$ is generated by the elements

$$\omega_{i_1}^h(x_1) * \cdots * \omega_{i_j}^h(x_j) := \lambda_h(\gamma_{i_1}(x_1)\cdots\gamma_{i_j}(x_j)) \tag{6.3}$$

with $i_k \geq 1$ for all k, and $\sum_k i_k = n$. The following description of the derived functors $L_i\Lambda^n$ is given in [18] Theorem 2.3.5:

$$L_i\Lambda^n(A) \simeq (\Omega_{i+1}(A)\otimes\Lambda^{n-i-1}(A))/\operatorname{Jac}_\Lambda . \tag{6.4}$$

Here $\operatorname{Jac}_\Lambda$ is the subgroup generated by the expressions

$$\sum_{k=1}^{j}\omega_{i_1}^h(x_1) * \cdots * \omega_{i_k-1}^h(x_k) * \cdots * \omega_{i_j}^h(x_j)\otimes x_k\wedge y_1\wedge\cdots y_{n-i-2}$$

for all h, with $\sum_{k=1}^j i_k = i+2$.

For the case of a $\mathbb{Z}/p$-vector space (p is a prime), the structure of $L_i\Lambda^n$ can be given as follows. Let A be a free abelian group. Then

$$\Omega_n(A\otimes\mathbb{Z}/p) \simeq \Gamma_n(A)\otimes\mathbb{Z}/p.$$

In this case, the derived functors 6.4 can be identified with images of the Koszul complex (see [7]):

$$\begin{aligned} L_i\Lambda^n(A\otimes\mathbb{Z}/p) = \operatorname{coker}\{&\Gamma_{i+2}(A)\otimes\Lambda^{n-i-2}(A)\otimes\mathbb{Z}/p \\ &\to \Gamma_{i+1}(A)\otimes\Lambda^{n-i-1}(A)\otimes\mathbb{Z}/p\} \\ &\subset \Gamma_i(A)\otimes\Lambda^{n-i}(A)\otimes\mathbb{Z}/p \subset A^{\otimes n}\otimes\mathbb{Z}/p. \end{aligned}$$

Now we observe that all differentials in the spectral sequence (6.1) are certain natural transformations of the type:

$$\text{polynomial functor of degree } < n \longrightarrow L_i\Lambda^n(A\otimes\mathbb{Z}/p) \subset A^{\otimes n}\otimes\mathbb{Z}/p\,.$$

By (2.7), any such natural transformation is zero. Hence, all differentials of this spectral sequence are zero maps.

6.2. Homotopy groups of $K(A,2)\vee K(B,2)$

Recall that, for a pair of path-connected spaces X, Y, there exists the following fibre sequences (see [1, 17])

$$\Sigma(\Omega X)\wedge(\Omega Y) \to X\vee Y \to X\times Y \tag{6.5}$$

and

$$\Sigma(\Omega X)\wedge(\Omega X) \to \Sigma\Omega X \to X\,. \tag{6.6}$$

For a pair of abelian groups A, B, taking $X = K(A,2), Y = K(B,2)$, we get the homotopy equivalence

$$\Sigma(\Omega X) \wedge (\Omega Y) \sim \Sigma K(A,1) \wedge K(B,1).$$

Therefore, for $B = \mathbb{Z}$, we have natural isomorphisms

$$\pi_n(K(A,2) \vee K(\mathbb{Z},2)) \simeq \pi_n(\Sigma^2 K(A,1)), \; n \geq 3. \tag{6.7}$$

Taking $A = B$, the fibre sequences (6.5) and (6.6) imply the isomorphisms

$$\pi_n(\Sigma K(A,1)) \cong \pi_n(K(A,2) \vee K(A,2)) \text{ for } n \geq 3. \tag{6.8}$$

Choose the simplest simplicial model for $K(A,1)$. Applying the inverse to the normalization functor in the sense of Dold–Kan to the complex $A[1]$, we obtain the abelian simplicial group $Q_*(A)$ with components

$$Q_i = (\underbrace{A \times \cdots \times A}_{i \text{ copies}}), \; i \geq 1$$

and the property $|Q_*(A)| \simeq K(A,1)$. The face and degeneracy maps in $Q_*(A)$ are standard, their structure follows from the construction of the inverse to the normalization functor.

The following fact follows directly from the Whitehead Theorem [29] (see [19], Proposition 4.3, for the simplicial version of the Whitehead Theorem): for a pair of abelian groups A, B, there is a homotopy equivalence

$$|Q_*(A) * Q_*(B)| \simeq \Omega(K(A,2) \vee K(B,2)).$$

Here $Q_*(A) * Q_*(B)$ is the free product of simplicial groups $Q_*(A)$ and $Q_*(B)$. That is, for $T_* := Q_*(A) * Q_*(B)$ the components are

$$T_n := (\underbrace{A \times \cdots \times A}_{n \text{ copies}}) * (\underbrace{B \times \cdots \times B}_{n \text{ copies}}).$$

Consider the lower central series filtration of this simplicial group:

$$T_* \supset \gamma_2(T_*) \supset \gamma_3(T_*) \supset \cdots.$$

This filtration defines the spectral sequence, which is natural on A and B, with initial terms

$$E^1_{p,q} = \pi_q(\gamma_p(T_*)/\gamma_{p+1}(T_*)).$$

There exists an increasing function $f : \mathbb{N} \to \mathbb{N}$, such that, for $m \geq 2$,

$$\pi_i(\gamma_m(T_*)/\gamma_{m+1}(T_*)) = 0, \; i < f(m). \tag{6.9}$$

The proof of 6.9 is routine. Here we give a general idea how to prove it.

Step 1. To reduce the problem to the connectivity for augmentation quotients. For a group G, denote its augmentation ideal $\ker\{\mathbb{Z}[G] \to \mathbb{Z}\}$ by $\Delta(G)$. Let G

be a group with torsion-free lower central quotients, then there is the following decomposition of augmentation quotients [27]:

$$\Delta^n(G)/\Delta^{n+1}(G) = \sum_{(a_1,\dots,a_n)} \bigotimes_{i=1}^{n} S^{a_i}(\gamma_i(G)/\gamma_{i+1}(G)),$$

where the sum runs over all non-negative $a_1, \dots, a_n$ such that $\sum_{i=1}^n ia_i = n$ (here $S^0(M) = \mathbb{Z}$ for an abelian group M). This "sum" presents a collection of graded factors of the functor Δ^n/Δ^{n+1}.

Step 2. To describe the structure of bifunctors

$$(A, B) \mapsto \Delta^n(A * B)/\Delta^{n+1}(A * B),\ n \geq 1.$$

There is a natural map

$$\bigoplus_{t_1+\dots+t_k=n} S^{t_1}(A_1) \otimes \cdots S^{t_k}(A_k) \to \Delta^n(A * B)/\Delta^{n+1}(A * B), \tag{6.10}$$

where all $A_i,\ i = 1, \dots, k$ are either A or B and A_i is not the same as A_{i+1} for all i. It is obvious that this map is surjective. Therefore, an abstract isomorphism between left and right sides of (6.10) implies that this natural transformation is an isomorphism. For example, there is a natural isomorphism

$$\Delta^3(A * B)/\Delta^4(A * B) \simeq (A \otimes B \otimes A) \oplus (B \otimes A \otimes B)$$
$$\oplus\, (S^2(A) \otimes B) \oplus (S^2(B) \otimes A) \oplus (A \otimes S^2(B)) \oplus (B \otimes S^2(A)) \oplus S^3(A) \oplus S^3(B).$$

Step 3. Use induction on degree of the functor and connectivity results for symmetric powers from [14].

Connectivity (6.9) implies that the spectral sequence converges:

$$E^1_{p,q} = \pi_q(\gamma_p(T_*)/\gamma_{p+1}(T_*)) \Rightarrow \pi_{q+1}(K(A,2) \vee K(B,2)),\ q \geq 2. \tag{6.11}$$

The idea is to analyze the bifunctors in the category of abelian groups:

$$\mathcal{F}_n: \quad (A, B) \mapsto \gamma_n(A * B)/\gamma_{n+1}(A * B),\ n \geq 1.$$

It is clear that

$$\mathcal{F}_1(A, B) = A \oplus B$$
$$\mathcal{F}_2(A, B) = A \otimes B.$$

The next terms have the structure

$$\mathcal{F}_3(A, B) = (A \otimes SP^2(B)) \oplus (B \otimes SP^2(A)).$$

The next functor $\mathcal{F}_4$ has graded pieces:

$$gr_1F_4(A, B) = \Lambda^2(A \otimes B),$$
$$gr_2F_4(A, B) = (A \otimes S^3(B)) \oplus (B \otimes S^3(A)),$$
$$gr_3F_4(A, B) = S^2(A) \otimes S^2(B).$$

The lower terms of this spectral sequence have the following structure:

$E^1_{p,q}$	$p=2$	3	4	5
$q=5$	0	$\mathrm{Tor}(A,\Omega_2(B))\oplus$ $\mathrm{Tor}(B,\Omega_2(A))$	$\pi_2\left(A\overset{L}{\otimes}L\Lambda^2(B)\right)\oplus$ $\Omega_2(A\otimes B)\oplus\mathrm{Tor}_2(A,B,\mathbb{Z}/2)\oplus$ $\pi_1\left(A\overset{L}{\otimes}L\Lambda^3(B)\oplus B\overset{L}{\otimes}L\Lambda^3(A)\right)$	*
4	0	$\pi_1\left(A\overset{L}{\otimes}L\Lambda^2(B)\right)\oplus$ $\pi_1\left(B\overset{L}{\otimes}L\Lambda^2(A)\right)$	$\Lambda^2(A\otimes B)\oplus\Lambda^2(A)\otimes\Lambda^2(B)\oplus$ $A\otimes\Lambda^3(B)\oplus B\otimes\Lambda^3(A)\oplus$ $\mathrm{Tor}_1(A,B,\mathbb{Z}/2)$	*
3	$\mathrm{Tor}(A,B)$	$\Lambda^2(A)\otimes B\oplus$ $\Lambda^2(B)\otimes A$	$A\otimes B\otimes\mathbb{Z}/2$	0
2	$A\otimes B$	0	0	0

It follows immediately that there is a natural isomorphism

$$\pi_3(K(A,2)\vee K(B,2))\simeq A\otimes B.$$

Now let us show that the differential

$$d_1:\ \pi_1\left(A\overset{L}{\otimes}L\Lambda^2(B)\oplus B\overset{L}{\otimes}L\Lambda^2(A)\right)\to A\otimes B\otimes\mathbb{Z}/2$$

is zero. We will use the fact that this is a natural transformation between bifunctors. It is enough to show that the vertical map

$$\begin{array}{ccc} A\otimes\Omega_2(B) & & \\ \downarrow & & \\ \pi_1\left(A\overset{L}{\otimes}L\Lambda^2(B)\right) & \longrightarrow & A\otimes B\otimes\mathbb{Z}/2 \\ \downarrow & & \\ \mathrm{Tor}(A,\Lambda^2(B)) & & \end{array}$$

induced by d^1 is zero, since A and B come to the picture in a symmetric way. First consider its restriction:

$$d^1|:A\otimes\Omega_2(B)\to A\otimes B\otimes\mathbb{Z}/2.$$

Since $A \otimes \Omega_2(B)$ and $A \otimes B \otimes \mathbb{Z}/2$ are additive right exact functors on A, it is enough to show that $d^1| = 0$ for $A = \mathbb{Z}$. Recall from 6.7 and [9] that

$$\pi_4(K(A,2) \vee K(\mathbb{Z},2)) = \pi_4(\Sigma^2 K(A,1)) = \widetilde{\otimes}^2(A)\,. \tag{6.12}$$

Therefore, $d^1|$ is zero. Hence the homomorphism d^1 can be extended to

$$\tilde{d}^1 : \mathrm{Tor}(A, \Lambda^2(B)) \to A \otimes B \otimes \mathbb{Z}/2.$$

Fixing A and looking at these functors as functors on B only, observe that, the natural map $B \oplus B \to B$, $(b_1, b_2) \mapsto b_1 + b_2$, induces an epimorphism from cross-effect to this functor:

$$\mathrm{Tor}(A, \Lambda^2(B|B)) = \mathrm{Tor}(A, B \otimes B) \to \mathrm{Tor}(A, \Lambda^2(B))\,.$$

Since the functor $A \otimes B \otimes \mathbb{Z}/2$ is additive on B, we see that $\tilde{d}^1$ is the zero map. Hence d^1 is zero.

Now the spectral sequence (6.11) implies that there are the following graded pieces of the fourth homotopy group of the wedge $K(A,2) \vee K(B,2)$ viewed as a bifunctor:

$$\begin{array}{ccccc} A \otimes B \otimes \mathbb{Z}/2 & \rightarrowtail & \text{some functor} & \twoheadrightarrow & (A \otimes \Lambda^2(B)) \oplus (B \otimes \Lambda^2(A)) \\ & & \downarrow & & \\ & & \pi_4(K(A,2) \vee K(B,2)) & & \\ & & \downarrow & & \\ & & \mathrm{Tor}(A,B)\,. & & \end{array} \tag{6.13}$$

Neither horizontal nor vertical sequence split naturally in (6.13). To see that the horizontal sequence does not split naturally, take $B = \mathbb{Z}$. In this case, the diagram (6.13) has the form

$$0 \to A \otimes \mathbb{Z}/2 \to \pi_4(K(A,2) \vee K(\mathbb{Z},2)) \to \Lambda^2(A) \to 0. \tag{6.14}$$

It follows from (6.12) that the sequence 6.14 does not split naturally.

Acknowledgement

The author thanks L. Breen for various discussions related to the subject of this paper and C. Vespa for important suggestions and corrections of some computations.

References

[1] W. Barkus and M. Barratt: On the homotopy classification of the extensions of the fixed map, *Trans. Amer. Math. Soc.* **88** (1958), 57–74.

[2] H.-J. Baues: *Homotopy type and homology*, Oxford Science Publications, Oxford, (1996).

[3] H.-J. Baues and P. Goerss: A homotopy operation spectral sequence for the computation of homotopy groups, *Topology* (2000).

[4] H.-J. Baues, W. Dreckmann, V. Franjou and T. Pirashvili: Foncteurs polynomiaux et founcteurs de Mackey non linéaires, *Bull. Soc. Math. France* **129** (2001), 237–257.

[5] H.-J. Baues and T. Pirashvili: Quadratic functors and one-connected two stage spaces, arxiv:math.0402250

[6] L. Breen: On the functorial homology of abelian groups, *J. Pure Appl. Alg.* **142** (1999), 199–237.

[7] L. Breen and R. Mikhailov: Derived functors of non-additive functors and homotopy theory, *Alg. Geom. Top.* **11** (2011), 327–415; arXiv: 0910.2817.

[8] L. Breen, R. Mikhailov and A. Touze: Derived functors of the divided power functors, to appear in *Geometry and Topology.*

[9] R. Brown and J.-L. Loday: Van Kampen theorems for diagrams of spaces, *Topology* **26** (1987), 311–335.

[10] G. Carlsson: A simplicial group construction for balanced products, *Topology* **23** (1985), 85–89.

[11] H. Cartan: Algèbres d'Eilenberg–Mac Lane et homotopie *Séminaire Cartan* **7** (1954/55), Secrétariat Mathématique.

[12] G.J. Decker, University of Chicago Ph.D. thesis (1974), available at: http://www.maths.abdn.ac.uk/ bensondj/html/archive/decker.html

[13] A. Dold: Homology of Symmetric Products and Other Functors of Complexes, *Annals of Math.*. **68** (1958), 54–80.

[14] A. Dold and D. Puppe: Homologie nicht-additiver Funktoren; Anwendugen. *Ann. Inst. Fourier* **11** (1961) 201–312.

[15] Yu. Drozd: On cubic functors, *Comm. Algebra* **31** (2003), 1147–1173.

[16] S. Eilenberg and S. Mac Lane: On the groups $H(\pi, n)$, II: Methods of computation, *Ann. Math.* **60**, (1954), 49–139.

[17] T. Ganea: A generalization of the homology and homotopy suspension, *Comm. Math. Helv.* **39** (1965), 295–322.

[18] F. Jean: Foncteurs dérivés de l'algébre symétrique: Application au calcul de certains groupes d'homologie fonctorielle des espaces $K(B, n)$, Ph.D. Thesis, University of Paris 13, 2002.

[19] D.M. Kan and W.P. Thurston: Every connected space has the homology of a $K(\pi, 1)$ *Topology* **15** (1976), 253–258.

[20] N. Kuhn: Generic representations of the finite general linear groups and the Steenrod algebra II, *K-theory* **8** (1994), 395–428.

[21] A. Liulevicius: A Theorem in Homological Algebra and Stable Homotopy of Projective Spaces, *Trans. Amer. Math. Soc.* **109** (1963), 540–552.

[22] S. Mac Lane: Triple torsion products and multiple Künneth formulas, *Math. Ann.* **140** (1960), 51–64.

[23] R. Mikhailov: On the homology of the dual de Rham complex, arxiv: 1001.2824.

[24] R. Mikhailov and J. Wu: On homotopy groups of the suspended classifying spaces, *Alg. Geom. Top.* **10** (2010), 565–625; arXiv: 0908.3580.

[25] D.G. Quillen: Spectral sequences of a double semi-simplicial group. *Topology* **5** (1966) 155–157.

[26] N. Roby, *Lois de polynômes et lois formelles en théorie des modules*, Annales Sci. de l'Éc. Norm. Sup, 3ème série, **80** (1953), 213–348.

[27] R. Sandling and K.-I. Tahara: Augmentation quotients of group rings and symmetric powers, Math. Proc. Camb. Math. Soc. **85** (1979), 247–252.

[28] J.H.C. Whitehead: A certain exact sequence, *Ann. Math.* **52** (1950), 51–110.

[29] J.H.C. Whitehead: On the asphericity of regions in a 3-sphere, *Fund. Math.* **32** (1939), 149–166.

Roman Mikhailov
Chebyshev Laboratory
St. Petersburg State University
14th Line, 29b
Saint Petersburg, 199178 Russia

and

St. Petersburg Department of
Steklov Mathematical Institute
e-mail: `rmikhailov@mail.ru`
URL: `http://chebyshev.spb.ru/roman_mikhailov`

Progress in Mathematics, Vol. 311, 99–149

Prerequisites of Homological Algebra

Antoine Touzé

Abstract. This text recalls the basic notions of homological algebra needed in the other parts of the book, in order to make them accessible to beginners. The aim of Sections 2 and 3 is to give a unified presentation of the various homology theories appearing in the book. In particular, we try to underline the links, the similarities and the differences between them. Section 4 is an introduction to spectral sequences, a technical tool which is used in all the other chapters.

Mathematics Subject Classification (2010). 20G10, 18G10, 55P65, 18G15, 20J06.

Keywords. Derived functors, group homology, functor categories, spectral sequences.

1. Introduction

Functors and short exact sequences

Let $\mathcal{A}$ be an abelian category, for example the category of modules over a ring R. To analyze the structure of an object $A \in \mathcal{A}$, we often cut it into smaller pieces: we find a subobject $A' \subset A$, and we study the properties of A' and of the quotient A/A'. Then we try to recover some information on A from the information we have on A' and A/A'.

Short exact sequences formalize the decomposition of an object into smaller pieces. They are diagrams in $\mathcal{A}$ of the following form, where (i) f is a monomorphism, (ii) the kernel of g is equal to the image of f and (iii) g is an epimorphism:

$$0 \to A' \xrightarrow{f} A \xrightarrow{g} A'' \to 0 .$$

The information we want to know about objects of $\mathcal{A}$ is often encoded by a functor from $\mathcal{A}$ to an other abelian category $\mathcal{B}$. For example, $\mathcal{A} = \mathrm{Ab}$ is the category of abelian groups and we want to determine the n-torsion part of an abelian group. In this case, F is the functor

$$\begin{array}{ccc} \mathrm{Ab} & \to & \mathbb{Z}/n\mathbb{Z}\text{-Mod} \\ A & \mapsto & {}_nA := \{a \in A\,;\, na = 0\} . \end{array}$$

A functor $F : \mathcal{A} \to \mathcal{B}$ is *exact* if it sends short exact sequences in $\mathcal{A}$ to short exact sequences in $\mathcal{B}$. If F is exact and if we know two of the objects $F(A')$, $F(A)$, $F(A'')$, then we have good chances to recover the third one.

Unfortunately, many interesting functors are not exact, but only semi-exact. For example, the n-torsion functor is only *left exact*, that is it sends a short exact sequence is to a sequence

$$0 \to {}_nA' \xrightarrow{{}_nf} {}_nA \xrightarrow{{}_ng} {}_nA''$$

where (i) ${}_nf$ is injective, (ii) the image of ${}_nf$ equals the kernel of ${}_ng$. But ${}_ng$ is not surjective in general as one easily sees by applying the n-torsion functor to the short exact sequence $0 \to \mathbb{Z} \xrightarrow{\times n} \mathbb{Z} \to \mathbb{Z}/n\mathbb{Z} \to 0$. In that case, to reconstruct one of the objects $F(A')$, $F(A)$ and $F(A'')$ from the two other ones, we need the theory of derived functors.

Derivation of semi-exact functors

By a semi-exact functor, we mean a functor which is left exact or right exact. When a functor F is left exact we can (if the category $\mathcal{A}$ has enough injectives, cf. Section 2) define its right derived functors R^iF, $i \geq 1$, which are designed so that the left exact sequence

$$0 \to F(A') \xrightarrow{F(f)} F(A) \xrightarrow{F(g)} F(A'')$$

fits into a *long exact sequence* (i.e., the kernel of each map equals the image of the preceding map)

$$0 \to F(A') \xrightarrow{F(f)} F(A) \xrightarrow{F(g)} F(A'') \xrightarrow{\delta} R^1F(A') \xrightarrow{R^1F(f)} R^1F(A) \xrightarrow{R^1F(g)} \cdots$$
$$\cdots \xrightarrow{\delta} R^iF(A') \xrightarrow{R^iF(f)} R^iF(A) \xrightarrow{R^iF(g)} R^iF(A'') \xrightarrow{\delta} \cdots$$

In particular, the cokernel of $F(g)$ equals the kernel of $R^1F(f)$, so the derived functors of F definitely help to recover one of the objects $F(A')$, $F(A)$ and $F(A'')$ from the two other ones.

Similarly, when a functor $G : \mathcal{A} \to \mathcal{B}$ is right exact (i.e., it sends short exact sequences $0 \to A' \to A \to A'' \to 0$ to right exact sequences $G(A') \to G(A) \to G(A'') \to 0$) we can define its left derived functors L^iG, $i \geq 1$ fitting into long exact sequences

$$\cdots L_1G(A) \to L_1G(A'') \to G(A') \to G(A) \to G(A'') \to 0\,.$$

(Co)homology theories as derived functors

As we will see in Section 2, the theory of derived functors of semi-exact functors provides a unified conceptual framework to study many (co)homology theories. For example the (co)homology of discrete groups or categories which appear in the lectures of A. Djament in this volume, or the cohomology of algebraic groups which appears in the lectures of W. van der Kallen in this volume.

Derived functors of non-additive functors

A functor $F : \mathcal{A} \to \mathcal{B}$ is *additive* if it commutes with direct sums, i.e., the canonical inclusions $A \hookrightarrow A \oplus A'$ and $A' \hookrightarrow A \oplus A'$ induce an isomorphism

$$F(A \oplus A') \simeq F(A) \oplus F(A') .$$

It is easy to show (see Exercise 2.3) that semi-exact functors are additive, as well as their derived functors. So the theory of derivation of semi-exact functors lives in the world of additive functors.

Unfortunately, there is a profusion of non-additive functors in algebra and topology. For example, the group ring $\mathbb{Z}A$ an abelian group A may be thought of as a non-additive functor $\mathrm{Ab} \to \mathrm{Ab}$. Similarly the homology $H_*(A,\mathbb{Z})$ of an abelian group A may be thought of as a non-additive functor with variable A.

For such general functors, Dold and Puppe provided a theory of derivation which we explain in Section 3. This theory provides a conceptual framework to study non-additive functors. For example, the homology of an abelian group A with coefficients in $\mathbb{Z}$ may be interpreted as derived functors of the group ring functor. Such derived functors appear in the lecture of R. Mikhailov in this volume.

Spectral sequences

Spectral sequences are a technical but essential tool in homological algebra. They play a crucial role to study and compute effectively derived functors. In Section 4 we provide an introduction to spectral sequences, with a focus on standard examples appearing in the remainder of the book.

2. Derived functors of semi-exact functors

2.1. Basic notions of homological algebra

2.1.1. Definitions related to complexes. Let $\mathcal{A}$ be an abelian category, e.g., the category R-Mod (resp. Mod-R) of left (resp. right) R-modules over a ring R.

A complex in $\mathcal{A}$ is a collection of objects $(C_i)_{i\in\mathbb{Z}}$ together with morphisms $d_C : C_i \to C_{i-1}$, satisfying $d_C \circ d_C = 0$. The ith homology object of C is the subquotient of C_i defined by:

$$H_i(C) = \mathrm{Ker}(d_C : C_i \to C_{i-1})/\mathrm{Im}(d_C : C_{i+1} \to C_i) .$$

If C and D are complexes, a map $f : C \to D$ is a collection of morphisms $f_i : C_i \to D_i$, satisfying $f_i \circ d = d \circ f_{i+1}$. It induces morphisms $H_i(f) : H_i(C) \to H_i(D)$ on the level of the homology. Two maps $f, g : C \to D$ are homotopic if there exists collection of morphisms $h_i : C_i \to D_{i+1}$, such that $f_i - g_i = d_D \circ h_i + h_{i-1} \circ d_C$ for all $i \geq 0$. Homotopic chain maps induce the same morphism in homology. A homotopy equivalence is a chain map $f : C \to D$ such that there exists $g : D \to C$ with $g \circ f$ and $f \circ g$ homotopic to the identity.

We will use the word *chain complex* to indicate a complex concentrated in nonnegative degrees, i.e., $C_i = 0$ for $i < 0$. We denote by $\mathrm{Ch}_{\geq 0}(\mathcal{A})$ the category of chain complexes and chain maps. We will use the word *cochain complex* to indicate a complex concentrated in nonpositive degrees, i.e., $C_i = 0$ for $i > 0$. By letting $C^i = C_{-i}$, we can see a cochain complex in a more familiar way as a collection of objects C^i, $i \geq 0$ with differentials $d_C : C^i \to C^{i+1}$ raising the degree by one. We denote by $\mathrm{Ch}^{\geq 0}(\mathcal{A})$ the category of cochain complexes.

For a nonnegative integer n, the n-fold suspension of a chain complex C is the chain complex $C[n]$ defined by $(C[n])_i = C_{i+n}$ and $d_{C[n]} = (-1)^n d_C$. This yields a functor $[n] : \mathrm{Ch}_{\geq 0}(\mathcal{A}) \to \mathrm{Ch}_{\geq 0}(\mathcal{A})$.

2.1.2. Additive functors. Let $F : \mathcal{A} \to \mathcal{B}$ be a functor between abelian categories. The functor F is called additive if it satisfies one of the following equivalent conditions.

(1) For all objects A, A' of $\mathcal{A}$ the canonical maps $A \to A \oplus A'$ and $A' \to A \oplus A'$ induce an isomorphism $F(A) \oplus F(A') \simeq F(A \oplus A')$.
(2) For all $f, g : A \to A'$, we have $F(f + g) = F(f) + F(g)$.

If F is an additive functor, then the functors $F : \mathrm{Ch}_{\geq 0}(\mathcal{A}) \to \mathrm{Ch}_{\geq 0}(\mathcal{B})$ and $F : \mathrm{Ch}^{\geq 0}(\mathcal{A}) \to \mathrm{Ch}^{\geq 0}(\mathcal{B})$ obtained by applying degreewise F preserve chain homotopies.

2.1.3. δ-functors

Definition 2.1. Let $\mathcal{A}$ and $\mathcal{B}$ be abelian categories. A *homological* δ-functor is a family of functors $G_i : \mathcal{A} \to \mathcal{B}$, $i \geq 0$, together with connecting morphisms $\delta_n : G_n(A'') \to G_{n-1}(A')$ defined for each exact sequence $0 \to A' \to A \to A'' \to 0$ such that:

(1) For each exact sequence $0 \to A' \to A \to A'' \to 0$ the diagram in $\mathcal{A}$ below is a long exact sequence (by convention, $G_i = 0$ if $i < 0$):
$$\cdots \xrightarrow{\delta_{i+1}} G_i(A') \to G_i(A) \to G_i(A'') \xrightarrow{\delta_i} G_{i-1}(A') \to \cdots .$$
(2) For each morphism of exact sequences, that is for each triple (f', f, f'') such that the following diagram commutes:
$$\begin{array}{ccccccccc} 0 & \longrightarrow & A' & \longrightarrow & A & \longrightarrow & A'' & \longrightarrow & 0 \\ & & \downarrow{\scriptstyle f'} & & \downarrow{\scriptstyle f} & & \downarrow{\scriptstyle f''} & & \\ 0 & \longrightarrow & B' & \longrightarrow & B & \longrightarrow & B'' & \longrightarrow & 0 \end{array},$$
we have a morphism of between the corresponding long exact sequences, that is the following diagram commutes:
$$\begin{array}{ccccccccc} \cdots & \longrightarrow & G_i(A') & \longrightarrow & G_i(A) & \longrightarrow & G_i(A'') & \longrightarrow & G_{i-1}(A') & \longrightarrow \cdots \\ & & \downarrow{\scriptstyle G_i(f')} & & \downarrow{\scriptstyle G_i(f)} & & \downarrow{\scriptstyle G_i(f'')} & & \downarrow{\scriptstyle G_{i-1}(f')} & \\ \cdots & \longrightarrow & G_i(B') & \longrightarrow & G_i(B) & \longrightarrow & G_i(B'') & \longrightarrow & G_{i-1}(B') & \longrightarrow \cdots \end{array}.$$

A morphism of δ-functors from $(G_i, \delta_i)_{i\geq 0}$ to $(H_i, \delta_i)_{i\geq 0}$ is a family of natural transformations $\theta_i : G_i \to H_i$, $i \geq 0$ such that for each short exact sequence in $\mathcal{A}$, the θ_i induce a morphism between the associated long exact sequences.

The name 'homological' δ-functor comes from the fact that the connecting morphisms decrease the degrees by one. Cohomological δ-functors are defined similarly with connecting morphisms raising the degrees by one.

Definition 2.2. A *cohomological δ-functor* is a family of functors $F^i : \mathcal{A} \to \mathcal{B}$, $i \geq 0$, equipped with connecting homomorphisms $\delta^i : F^i(A'') \to F^{i+1}(A')$ defined for each short exact sequence $0 \to A' \to A \to A'' \to 0$, such that we have long exact sequences:

$$0 \to F^0(A') \to F^0(A) \to F^0(A'') \xrightarrow{\delta^0} F^1(A') \to \cdots$$

and such that each morphism of short exact sequences induces a morphism between the corresponding long exact sequences.

The following exercise shows that the functors appearing as a component of a (co)homological δ-functor are additive.

Exercise 2.3. Let $G : \mathcal{A} \to \mathcal{B}$ be a functor between abelian categories, such that for all short exact sequences $0 \to A' \to A \to A'' \to 0$ the image of $G(A') \to G(A)$ is equal to the kernel of $G(A) \to G(A'')$. Show that G is additive. (Hint: use the split exact sequence $0 \to A \to A \oplus B \to B \to 0$.)

2.1.4. Projective resolutions

Definition 2.4. Let $\mathcal{A}$ be an abelian category. An object P of $\mathcal{A}$ is *projective* if the following functor is exact:

$$\begin{array}{rccc} \mathrm{Hom}_{\mathcal{A}}(P, -) : & \mathcal{A} & \to & \mathrm{Ab} \\ & A & \mapsto & \mathrm{Hom}_{\mathcal{A}}(P, A)\,. \end{array}$$

Exercise 2.5. Show that the projective objects of the category R-Mod of modules over a ring R are the direct summands of free R-modules. Show that there might be projective R-modules which are not free (example: consider the ring $\mathbb{Z} \times \mathbb{Z}$).

Definition 2.6. An abelian category $\mathcal{A}$ has *enough projectives* if all objects A admit a projective resolution. This means that for all A, there exists a chain complex P^A of projective objects such that $H_i(P^A) = A$ if $i = 0$ and zero otherwise.

Exercise 2.7. Let R be a ring. Show that R-Mod and Mod-R have enough projectives.

It is not true that all abelian categories have enough projectives. The category of rational GL_n-modules in Section 2.6 provides a counter-example. To compare the various projective resolutions which might occur in an abelian category $\mathcal{A}$, we can use the following fundamental lemma, whose proof is left as an exercise.

Lemma 2.8. *Let A, B be objects of an abelian category $\mathcal{A}$. Let P^A and P^B be projective resolutions of A and B in $\mathcal{A}$. Then for all $f : A \to B$, there exists a chain map $\overline{f} : P^A \to P^B$ such that $H_0(\overline{f}) = f$. Such a chain map is unique up to homotopy. In particular a projective resolution of A is unique up to a homotopy equivalence.*

2.1.5. Injective coresolutions

Definition 2.9. Let $\mathcal{A}$ be an abelian category. An object J of $\mathcal{A}$ is *injective* if the following contravariant functor is exact:

$$\begin{array}{rccc} \mathrm{Hom}_{\mathcal{A}}(-, J) : & \mathcal{A} & \to & \mathrm{Ab} \\ & A & \mapsto & \mathrm{Hom}_{\mathcal{A}}(A, J) \end{array}$$

Exercise 2.10. Show that an object of $\mathcal{A}$ is injective if and only if it is projective in the opposite category[1].

Definition 2.11. An abelian category $\mathcal{A}$ has *enough injectives* if all objects A admit an injective coresolution. This means that for all A, there exists a cochain complex J_A of injective objects, such that $H^i(J_A) = A$ if $i = 0$ and zero otherwise.

Not all abelian categories have enough injectives. With a little work, one can prove:

Proposition 2.12. *The categories R-*Mod *(resp.* Mod-R*) of left (resp. right) modules over a ring R have enough injectives.*

Sketch of proof. We do the proof for R-Mod. The proof decomposes in several steps.

1. We first observe that to prove that an abelian category $\mathcal{A}$ has enough injective objects, it suffices to prove that for all objects A, we can find an injective object and an injective map $A \hookrightarrow J$.
2. **$\mathbb{Q}/\mathbb{Z}$ is injective in** Ab. An abelian group G is *divisible* if for all $x \in G$ and all $n \in \mathbb{Z} \setminus \{0\}$, there exists $x' \in G$ such that nx'.

   To prove that $\mathbb{Q}/\mathbb{Z}$ is injective, we first prove that being injective is equivalent to being divisible. Then we check that $\mathbb{Q}/\mathbb{Z}$ is divisible.
3. **$\mathrm{Hom}_{\mathbb{Z}}(R, \mathbb{Q}/\mathbb{Z})$ is injective in R-mod.** Take the abelian group $\mathrm{Hom}_{\mathbb{Z}}(R, \mathbb{Q}/\mathbb{Z})$ as an R-module, with action of R on a function $f : R \to \mathbb{Q}/\mathbb{Z}$ given by $rf(x) := f(xr)$. There is an isomorphism of abelian groups, natural with respect to the R-module M

$$\mathrm{Hom}_R(M, \mathrm{Hom}_{\mathbb{Z}}(R, \mathbb{Q}/\mathbb{Z})) \simeq \mathrm{Hom}_{\mathbb{Z}}(M, \mathbb{Q}/\mathbb{Z}) \ . \quad (*)$$

   In particular, since $\mathbb{Q}/\mathbb{Z}$ is injective, $\mathrm{Hom}_{\mathbb{Z}}(R, \mathbb{Q}/\mathbb{Z})$ is also injective.
4. We observe that an arbitrary product of injective R-modules is still injective.

[1]the opposite category of $\mathcal{A}$ is the category $\mathcal{A}^{\mathrm{op}}$ with the same objects as $\mathcal{A}$, with

$$\mathrm{Hom}_{\mathcal{A}^{\mathrm{op}}}(A, B) := \mathrm{Hom}_{\mathcal{A}}(B, A),$$

and the composition in $\mathcal{A}^{\mathrm{op}}$ is defined by $f \circ g := g \circ_{\mathcal{A}} f$, where $\circ_{\mathcal{A}}$ denotes the composition in $\mathcal{A}$.

5. Finally, let M be an R-module. Let J the injective R-module obtained as the product of copies of $J_R = \mathrm{Hom}_{\mathbb{Z}}(R, \mathbb{Q}/\mathbb{Z})$, indexed by the set $\mathcal{M} = \mathrm{Hom}_R(M, J_R)$. There is a canonical R-linear map $\phi : M \to J$.
   To be more specific, the coordinate of ϕ indexed by $f \in \mathcal{M}$ is the map $\phi_f : M \to J_R$ which sends $m \in M$ to $f(m)$. To prove that ϕ is an injection, it suffices to find for all $m \in M$ an element $f \in \mathcal{M}$ such that $f(m) \neq 0$ or equivalently (cf. isomorphism $(*)$) a morphism of abelian groups $\overline{f} : M \to \mathbb{Q}/\mathbb{Z}$ such that $\overline{f}(m) \neq 0$. Such an $\overline{f}$ can be produced using the injectivity of $\mathbb{Q}/\mathbb{Z}$. □

The following fundamental lemma can be formally deduced from Lemma 2.8 and Exercise 2.10.

Lemma 2.13. *Let A, B be objects of an abelian category $\mathcal{A}$. Let J_A and J_B be injective resolutions of A and B in $\mathcal{A}$. Then for all $f : A \to B$, there exists a chain map $\overline{f} : J_A \to J_B$ such that $H^0(\overline{f}) = f$. Such a chain map is unique up to homotopy.*

2.2. Derivation of semi-exact functors

2.2.1. Derivation of right exact functors. If $(G_i, \delta_i)_{i\geq 0}$ is a homological δ-functor, the 'end' of the long exact sequences

$$\cdots \to G_0(A') \to G_0(A) \to G_0(A'') \to 0$$

imply that G_0 is a right exact functor. In particular, the assignment $(G_i, \delta_i)_{i\geq 0} \mapsto G_0$ defines a functor from the category of homological δ-functors, to the category of right exact functors (with natural transformations as morphisms):

$$\left\{ \begin{array}{c} \text{homological } \delta\text{-functors} \\ \text{from } \mathcal{A} \text{ to } \mathcal{B} \end{array} \right\} \to \left\{ \begin{array}{c} \text{right exact functors} \\ \text{from } \mathcal{A} \text{ to } \mathcal{B} \end{array} \right\}.$$

Derivation of functors yields an operation going in the other way, which enables (in good cases, cf. condition (ii) in Theorem 2.14) to reconstruct the homological functor $(G_i, \delta_i)_{i\geq 0}$ from G_0.

Theorem 2.14. *Let $\mathcal{A}$ and $\mathcal{B}$ be two abelian categories, and let $G : \mathcal{A} \to \mathcal{B}$ be a right exact functor. Assume that $\mathcal{A}$ has enough projectives. There exists a homological δ-functor $(L_iG, \delta_i) : \mathcal{A} \to \mathcal{B}$ such that:*

(i) $L_0G = G$,
(ii) *for all $i > 0$, $L_iG(P) = 0$ if P is projective.*

Such a δ-functor is unique up to isomorphism. Moreover, if G and G' are right exact functors, a natural transformation $\theta : G \to G'$ extends uniquely into a morphism of homological δ-functors $\theta_i : L_iG \to L_iG'$ such that $\theta_0 = \theta$.

Definition 2.15. The functors L_iG, $i \geq 0$ from Theorem 2.14 are called the left derived functors of G.

We now outline the main ideas of the proof of Theorem 2.14.

1. If $(G_i,\delta_i)_{i\ge 0}$ and $(G'_i,\delta'_i)_{i\ge 0}$ are homological δ-functors satisfying condition (ii), then a natural transformation $\theta : G_0 \to G'_0$ may be uniquely extended into a morphism of δ-functors θ_i, such that $\theta_0 = \theta$.

   This results from the following 'dimension shifting argument'. Let $\theta_i : G_i \to G'_i$ be a morphism of δ-functors. For all $M \in \mathcal{A}$ we can find a projective P and an epimorphism $P \twoheadrightarrow M$. Let K be the kernel of this epimorphism. The long exact sequences of G_i and G'_i provide commutative diagrams:

$$\begin{array}{ccc} G_1(M) & \overset{\delta_1}{\hookrightarrow} & G_0(K_M) \longrightarrow \cdots \\ \downarrow{\scriptstyle \theta_{1\,M}} & & \downarrow{\scriptstyle \theta_{0\,K}} \\ G_1(M) & \overset{\delta'_1}{\hookrightarrow} & G_0(K_M) \longrightarrow \cdots \end{array}\quad \text{and}\quad \begin{array}{ccc} G_{i+1}(M) & \xrightarrow[\simeq]{\delta_{i+1}} & G_i(M) \\ \downarrow{\scriptstyle \theta_{i+1\,M}} & & \downarrow{\scriptstyle \theta_{i\,K}} \\ G'_{i+1}(M) & \xrightarrow[\simeq]{\delta'_{i+1}} & G_i(M) \end{array}\quad i \ge 1.$$

   So the natural transformations θ_i, $i \ge 0$ are completely determined by θ_0. Whence the uniqueness of the extension of θ.

   For the existence, one checks that taking $\theta_{1\,M}$ as the restriction of θ_K to $G_1(M)$ and $\theta_{i+1\,M} = (\delta'_{i+1})^{-1} \circ \theta_{i\,K} \circ \delta_{i+1}$ provides a well-defined morphism of δ-functors extending θ.
2. The uniqueness of $(L_iG,\delta_i)_{i\ge 0}$ satisfying (i) and (ii) follows the existence and uniqueness of the extension of the natural transformation $\mathrm{Id} : G \to G$.
3. The construction of a homological δ-functor $(L_iG,\delta_i)_{i\ge 0}$ satisfying (i) and (ii) is achieved by the following recipe. We first fix for each object A a projective resolution P^A. Then L_iG is defined on objects by:

$$L_iG(A) = H_i(G(P^A))\,.$$

   To define L_iG on morphisms, we use the fundamental Lemma 2.8. Each map $f : A \to B$ induces a chain map $\overline{f} : P^A \to P^B$, and we define $L_iG(f)$ as the map

$$H_i(G(\overline{f})) : H_i(G(P^A)) \to H_i(G(P^B))\,.$$

   Since $\overline{f}$ is unique up to homotopy and G is additive (see Exercise 2.3), $G(\overline{f})$ is unique up to homotopy, hence different choices of $\overline{f}$ induce the same morphism on homology. So $H_i(G(\overline{f}))$ is well defined. To finish the proof of Theorem 2.14 it remains to check that the functors L_iG actually form a δ-functor. This verification is rather long so we refer the reader to the classical references [ML, Wei].

Exercise 2.16. Assume that $\mathcal{A}$ has enough projectives. Prove that a right exact functor G is exact if and only if its left derived functors L_iG are zero for $i > 1$.

Exercise 2.17. Find two non-isomorphic δ-functors $(G_i,\delta_i)_{i\ge 0}$ and $(G'_i,\delta'_i)_{i\ge 0}$ such that $G_0 \simeq G'_0$.

2.2.2. Derivation of left exact functors. Derivation of right exact functors has the following analogue in the case of left exact functors.

Theorem 2.18. *Let $\mathcal{A}$ and $\mathcal{B}$ be two abelian categories, and let $F : \mathcal{A} \to \mathcal{B}$ be a left exact functor. Assume that $\mathcal{A}$ has enough injectives. There exists a cohomological δ-functor $(R^iG, \delta^i) : \mathcal{A} \to \mathcal{B}$ such that:*
(i) $R^0F = F$,
(ii) *for all $i > 0$, $R^iF(J) = 0$ if J is injective.*
Such a δ-functor is unique up to isomorphism. Moreover, if F and F' are left exact functors, a natural transformation $\theta : F \to F'$ extends uniquely into a morphism of δ-functors $\theta^i : R^iF \to R^iF'$ such that $\theta^0 = \theta$.

The proof of Theorem 2.18 is completely similar to the case of right exact functors, so we omit it. Let us just mention that the derived functor R^iF sends an object A to the homology group $H^i(F(J_A))$, and for $f : A \to B$, the morphism $R^i(f)$ is equal to $H^i(F(\overline{f}))$, where $\overline{f} : J_A \to J_B$ is a lifting of f to the injective coresolutions.

Exercise 2.19. Prove that a left exact functor $F : \mathcal{A} \to \mathcal{B}$ is equivalent to a right exact functor $\mathcal{A}^{\mathrm{op}} \to \mathcal{B}^{\mathrm{op}}$, and that a cohomological δ-functor $(F^i, \delta^i)_{i\geq 0} : \mathcal{A} \to \mathcal{B}$ is equivalent to an homological δ-functor $\mathcal{A}^{\mathrm{op}} \to \mathcal{B}^{\mathrm{op}}$. Deduce Theorem 2.18 from the statement of Theorem 2.14.

2.3. Ext and Tor

The most common examples of derived functors are the functors Tor and Ext. Basic references for this section are [ML, III] and [Wei, Chap. 3].

2.3.1. Ext. Let $\mathcal{A}$ be an abelian category. Assume that $\mathcal{A}$ is enriched over a commutative ring $\Bbbk$ (one also says that A is $\Bbbk$-linear). This means that Hom groups in $\mathcal{A}$ are $\Bbbk$-modules, and composition in $\mathcal{A}$ is $\Bbbk$-bilinear. For example, the category of left (or right) modules over a $\Bbbk$-algebra R are of that kind. And abelian categories are by definition enriched over $\mathbb{Z}$. Then for objects M, N in $\mathcal{A}$ we have left exact functors:

$$\mathrm{Hom}_{\mathcal{A}}(M,-) : \mathcal{A} \to \Bbbk\text{-Mod}\,, \qquad \mathrm{Hom}_{\mathcal{A}}(-,N) : \mathcal{A}^{\mathrm{op}} \to \Bbbk\text{-Mod}\,.$$

If $\mathcal{A}$ has enough injectives, we define Ext-functors by the formula:

$$\mathrm{Ext}^i_{\mathcal{A}}(M,-) = R^i(\mathrm{Hom}_{\mathcal{A}}(M,-))\,.$$

If $\mathcal{A}^{\mathrm{op}}$ has enough injectives (this is equivalent to the fact that $\mathcal{A}$ has enough projectives) then we define Ext-functors by the formula

$$\mathrm{Ext}^i_{\mathcal{A}}(M,-) = R^i(\mathrm{Hom}_{\mathcal{A}}(-,N))\,.$$

Remark 2.20. Assume that $\mathcal{A}$ has enough injectives and enough projectives. Then the notation $\mathrm{Ext}^i_{\mathcal{A}}(M,N)$ might have two different meanings: either the value of $R^i(\mathrm{Hom}_{\mathcal{A}}(M,-))$ on N or the value of $R^i(\mathrm{Hom}_{\mathcal{A}}(-,N))$ on M. It can be proved that these two definitions coincide. That is, $\mathrm{Ext}^i_{\mathcal{A}}(M,N)$ can be indifferently computed as the ith homology of the cochain complex $\mathrm{Hom}_{\mathcal{A}}(P^M,N)$ where P^M is a projective resolution of M or as the ith homology of the cochain complex $\mathrm{Hom}_{\mathcal{A}}(M,J_N)$ where J_N is an injective resolution of N.

The notation Ext is the abbreviation of 'Extension groups'. If A, B are objects of $\mathcal{A}$, an extension of A by B in $\mathcal{A}$ is a short exact sequence $0 \to B \to C \to A \to 0$. Two extensions are isomorphic if we have a commutative diagram (in this case f is automatically an isomorphism):

$$\begin{array}{ccccccccc} 0 & \longrightarrow & A & \longrightarrow & C & \longrightarrow & B & \longrightarrow & 0 \\ & & \| & & \downarrow f & & \| & & \\ 0 & \longrightarrow & A & \longrightarrow & C' & \longrightarrow & B & \longrightarrow & 0 . \end{array}$$

An extension is trivial when it is isomorphic to $0 \to A \to A \oplus B \to B \to A$. We let $\mathcal{E}(A, B)$ be the set of isomorphism classes of extensions of A by B, pointed by the class of trivial extensions. The following proposition is the reason for the name 'Ext' (such an approach of Exts is used in R. Mikhailov's lectures).

Proposition 2.21. *There is a bijection of pointed sets (on the left-hand side* $\mathrm{Ext}^1_{\mathcal{A}}(A, B)$ *is pointed by* 0*):*

$$\mathrm{Ext}^1_{\mathcal{A}}(A, B) \simeq \mathcal{E}(A, B) .$$

Actually much more can be said: we can describe explicitly the abelian group structure in terms of operations on $\mathcal{E}(A, B)$, and higher Ext also have analogous interpretations. We refer the reader to [ML, III] for more details on these topics (and for the proof of Proposition 2.21).

Exercise 2.22. Let $\mathcal{A}$, $\mathcal{B}$ be abelian categories with enough projectives and enough injectives. Let $G : \mathcal{B} \to \mathcal{A}$ be a right adjoint to $F : \mathcal{A} \to \mathcal{B}$. Recall that this means that there is a bijection, natural in x, y:

$$\mathrm{Hom}_{\mathcal{B}}(F(x), y) \simeq \mathrm{Hom}_{\mathcal{A}}(x, G(y)) .$$

1. Show that F is right exact and that G is left exact.
2. Show that G is exact if and only if F preserves the projectives (and similarly F is exact if and only if G preserves the injectives).
3. Show that F and G are both exact if and only if there are isomorphisms $\mathrm{Ext}^*_{\mathcal{B}}(F(x), y) \simeq \mathrm{Ext}^*_{\mathcal{A}}(x, G(y))$ for all x, y.

2.3.2. Tor. In this section, we fix a commutative ring $\Bbbk$ and a $\Bbbk$-algebra R (if $\Bbbk = \mathbb{Z}$, a $\Bbbk$-algebra is nothing but an ordinary ring). If M is a right R-module and N is a left R-module, the tensor product $M \otimes_R N$ is the $\Bbbk$-module generated by the symbols $m \otimes n$, $m \in M$, $n \in N$, with the following relations:

$$m \otimes n + m' \otimes n = (m + m') \otimes n ,$$

$$m \otimes n + m \otimes n' = (m + m') \otimes n ,$$

$$\lambda(m \otimes n) = (m\lambda) \otimes n = m \otimes (\lambda n) \text{ for } \lambda \in \Bbbk ,$$

$$(mr) \otimes n = m \otimes (rn) \text{ for } r \in R .$$

Tensor product over the $\Bbbk$-algebra R yields functors:

$$M \otimes_R : R\text{-Mod} \to \Bbbk\text{-Mod} , \qquad \otimes_R N : \text{Mod-}R \to \Bbbk\text{-Mod} .$$

Exercise 2.23. Prove that we have isomorphisms, natural with respect to $M \in$ Mod-R, $N \in R$-Mod, and $K \in \Bbbk$-Mod:

$$\mathrm{Hom}_{\Bbbk}(M \otimes_R N, Q) \simeq \mathrm{Hom}_{\text{Mod-}R}(M, \mathrm{Hom}_{\Bbbk}(N, Q)) ,$$
$$\mathrm{Hom}_{\Bbbk}(M \otimes_R N, Q) \simeq \mathrm{Hom}_{R\text{-Mod}}(N, \mathrm{Hom}_{\Bbbk}(M, Q)) .$$

Deduce[2] from these isomorphisms the following properties of the functors $M\otimes_R$ and $\otimes_R N$: (i) they are right exact, (ii) they commute with arbitrary sums, and (iii) if M (resp. N) is projective, then $M\otimes_R$ (resp. $\otimes_R N$) is exact.

By Exercise 2.23, $M\otimes_R$ and $\otimes_R N$ are right exact. Tor-functors are defined by the following formulas (recall the category of left or right R-modules has enough projectives so that left derived functors are well defined).

$$\mathrm{Tor}_i^R(-, N) = L_i(\otimes_R N) , \quad \mathrm{Tor}_i^R(M, -) = L_i(M\otimes_R) .$$

Remark 2.24. The notation $\mathrm{Tor}_i^R(M, N)$ has two different meanings: on the one hand it is the value on M of $L_i(\otimes_R N)$ and on the other hand it is the value of $L_i(M\otimes_R)$ on N. However, we shall see in Example 4.22 that they coincide. Thus, the $\Bbbk$-module $\mathrm{Tor}_i^R(M, N)$ can be indifferently be computed as the ith homology of the complex $P^M \otimes_R N$ or of the complex $N \otimes_R P^N$, where P^X denotes a projective resolution of X.

The name 'Tor' is the abbreviation of 'Torsion', and it is justified by the case of abelian groups, as the following exercise shows it.

Exercise 2.25. Let A and B be abelian groups. Show that $\mathrm{Tor}_i^{\mathbb{Z}}(A, B) = 0$ for $i > 1$. If $B = \mathbb{Z}/n\mathbb{Z}$, show that $\mathrm{Tor}_1^{\mathbb{Z}}(A, \mathbb{Z}/n\mathbb{Z})$ is the functor sending an abelian group A to its n-torsion part ${}_nA = \{a \in A \ : na = 0\}$.

The following exercise explains why so many examples of left derived functors can be interpreted as functors $\mathrm{Tor}_*^R(M, -)$.

Exercise 2.26. Let R be a $\Bbbk$-algebra, and let $G : R\text{-Mod} \to \Bbbk\text{-Mod}$ be a right exact functor, commuting with arbitrary sums.

1. Show that functoriality endows the $\Bbbk$-module $G(R)$ with the structure of a right R-module.
2. Show that for all free R-modules F, there is an isomorphism of $\Bbbk$-modules, natural with respect to F: $G(R) \otimes_R F \simeq G(F)$.
3. Deduce that for all $i \geq 0$, $L_iG(-) \simeq \mathrm{Tor}_i^R(G(R), -)$.

2.3.3. Bar complexes. Let R be a $\Bbbk$-algebra, let $M \in$ Mod-R and $N \in R$-Mod. The double-sided bar complex $B(M, R, N)$ is the complex of $\Bbbk$-modules defined as follows. As a graded $\Bbbk$-module, we have (tensor products are taken over $\Bbbk$):

$$B(M, R, N)_k := M \otimes R^{\otimes k} \otimes N .$$

An element $m \otimes x_1 \otimes \cdots \otimes x_k \otimes n \in B(M, R, N)_k$ is denoted by $m[x_1| \cdots |x_k]n$, and an element of $B(M, R, N)_0$ is denoted by $m[\,]n$. This handy notation is the origin

[2]This question requires the Yoneda lemma, see Section 2.5.2.

of the name 'bar resolution'. The differential $d : B(M,R,N)_k \to B(M,R,N)_{k-1}$ is the $\Bbbk$-linear map defined by

$$\begin{aligned} d(m[x_1,\dots,x_k]n) = {} & mx_1[x_2|\cdots|x_k]n \\ & + \textstyle\sum_{i=1}^{k-1}(-1)^i m[x_1|\cdots|x_ix_{i+1}|\cdots|x_k]n \\ & + (-1)^n m[x_1|\cdots|x_{k-1}]x_kn\,. \end{aligned}$$

If $M = R$, then $B(R,R,N)$ becomes a complex in the category of R-modules. The action of R on $B(R,R,N)_k$ is given by the formula:

$$x\cdot(m[x_1|\cdots|x_k]n) := (xm)[x_1|\cdots|x_k]n\,.$$

Proposition 2.27. *$B(R,R,N)$ is a resolution of N in the category of left R-modules. Furthermore, if R and N are projective as $\Bbbk$-modules, then it is a projective resolution of N in R-*Mod.

Proof. We have already seen that $B(R,R,N)$ is a complex of R-modules. Furthermore, if N and R are projective as $\Bbbk$-modules, then the R-modules $B(R,R,N)_k$ are projective R-modules. So, to prove Proposition 2.27 we want to prove that $H_i(B(R,R,N)) = N$ if $i = 0$ and zero otherwise. Let us consider N as a complex of R-modules concentrated in degree zero. The multiplication

$$\begin{array}{rccc} \lambda: & R\otimes N = B(R,R,N)_0 & \to & N \\ & x\otimes n & \mapsto & xn \end{array}$$

yields a morphism of complexes of R-modules $B(R,R,N) \xrightarrow{\lambda} N$. So, to prove Proposition 2.27, it suffices to prove that $H_*(\lambda)$ is an isomorphism.

To prove this, we forget the action of R and consider λ as a morphism of complexes of $\Bbbk$-modules. We are actually going to prove that λ is a homotopy equivalence of complexes of $\Bbbk$-modules. The inverse of λ is the map $\eta : N \to B(R,R,N)$ which sends an element $n \in N$ to $1\otimes N \in R\otimes N = B(R,R,N)_0$ (Notice that η is not R-linear, it is only a morphism of complexes of $\Bbbk$-modules.) Then $\epsilon\circ\eta = \mathrm{Id}$, and $\eta\circ\epsilon$ is homotopic to the identity via the homotopy h defined by $h(x[x_1|\cdots|x_k]n) = 1[x|x_1|\cdots|x_k]n$. □

Similarly, we can prove that if R and M are projective as $\Bbbk$-modules, then $B(M,R,R)$ is a projective resolution of M in Mod-R. As a consequence of Proposition 2.27, we obtain a nice explicit complex to compute Tors.

Corollary 2.28. *Let R be a $\Bbbk$-algebra, let $M \in$* Mod-R *and $N \in R$-*Mod. *Assume that R, and at least one of the two R-modules M and N are projective as $\Bbbk$-modules. Then the homology of $B(M,R,N)$ equals* $\mathrm{Tor}^R_*(M,N)$.

Proof. Assume that R and N are projective $\Bbbk$-modules. Then $B(R,R,N)$ is a projective resolution of N. Hence $M\otimes_R B(R,R,N)$ computes $\mathrm{Tor}^R_*(M,N)$. But $M\otimes_R B(R,R,N)$ equals $B(M,R,N)$. □

We can also use bar complexes to compute extension groups.

Corollary 2.29. *Let R be a $\Bbbk$-algebra, let M, N be left R-modules. Assume that R and M are projective as $\Bbbk$-modules. Then $\mathrm{Ext}^*_R(M,N)$ equals the homology of the complex $\mathrm{Hom}_R(B(R,R,M),N)$.*

2.4. Homology and cohomology of discrete groups

As basic references for the (co)homology of groups, the reader can consult [Wei, Chap. 6], [ML, IV], [Br] or [Ben, Chap. 2].

2.4.1. Definitions. Let G be a group, and let $\Bbbk$ be a commutative ring. A $\Bbbk$-linear representation of G is a $\Bbbk$-module M, equipped with an action of G by $\Bbbk$-linear morphisms. A morphism of representations $f : M \to N$ is a $\Bbbk$-linear map which commutes with the action of G, i.e., $f(gm) = gf(m)$ for all $m \in M$ and all $g \in G$.

Let $\Bbbk G$ be the group algebra[3] of G over $\Bbbk$. Then the category of $\Bbbk$-linear representations of G is isomorphic to the category of (left) $\Bbbk G$-modules. In particular, it has enough injectives and projectives.

Let $\Bbbk^{\mathrm{triv}}$ be the trivial $\Bbbk G$-module, that is the $\Bbbk$-module $\Bbbk$, acted on by G by $gx = x$. If M is a representation of G, its fixed points (or invariants) under the action of G is the $\Bbbk$-module M^G defined by:

$$M^G = \{m \in M\ , gm = m\ \forall g \in G\} = \mathrm{Hom}_{\Bbbk G}(\Bbbk^{\mathrm{triv}}, M)\ .$$

Its coinvariants is the $\Bbbk$-module M_G defined by:

$$M_G = M/\langle gm - m\ ,\ m \in M, g \in G\rangle = \Bbbk^{\mathrm{triv}} \otimes_{\Bbbk G} M\ .$$

The (co)homology of G is defined by deriving the functor of (co)invariants.

$$H^i(G,-) = R^i(-^G) = \mathrm{Ext}^i_{\Bbbk G}(\Bbbk^{\mathrm{triv}},-)\ ,$$
$$H_i(G,-) = L_i(-_G) = \mathrm{Tor}^{\Bbbk G}_i(\Bbbk^{\mathrm{triv}},-)\ .$$

We don't mention the ring $\Bbbk$ in the notations $H^i(G,M)$ and $H_i(G,M)$. Since a $\Bbbk G$-module is also a $\mathbb{Z}G$-module, these notations have two interpretations, e.g., $H^i(G,M)$ means $\mathrm{Ext}^i_{\Bbbk G}(\Bbbk^{\mathrm{triv}},M)$ as well as $\mathrm{Ext}^i_{\mathbb{Z}G}(\mathbb{Z}^{\mathrm{triv}},M)$. The following exercise shows that these two interpretations coincide.

Exercise 2.30. If M is a $\mathbb{Z}G$-module and N is a $\Bbbk G$-module, there is a canonical structure of $\Bbbk G$-module on $\Bbbk \otimes_{\mathbb{Z}} M$, and a canonical structure of $\Bbbk$-module on $\mathrm{Hom}_{\mathbb{Z}G}(M,N)$. Prove that there is an isomorphism of $\Bbbk$-modules

$$\mathrm{Hom}_{\mathbb{Z}G}(M,N) \simeq \mathrm{Hom}_{\Bbbk G}(\Bbbk \otimes_{\mathbb{Z}} M, N)\ .$$

Recall the bar complex from Section 2.3.3. Show that $\Bbbk \otimes_{\mathbb{Z}} B(\mathbb{Z}G,\mathbb{Z}G,\mathbb{Z}^{\mathrm{triv}})$ is isomorphic to $B(\Bbbk G,\Bbbk G,\Bbbk^{\mathrm{triv}})$ as a complex of $\Bbbk G$-modules. Deduce that there is an isomorphism of $\Bbbk$-modules $\mathrm{Ext}^i_{\mathbb{Z}G}(\mathbb{Z}^{\mathrm{triv}},M) \simeq \mathrm{Ext}^i_{\Bbbk G}(\Bbbk^{\mathrm{triv}},M)$. Similarly, prove the isomorphism $\mathrm{Tor}^{\mathbb{Z}G}_i(\mathbb{Z}^{\mathrm{triv}},M) \simeq \mathrm{Tor}^{\Bbbk G}_i(\Bbbk^{\mathrm{triv}},M)$.

---

[3]The group algebra $\Bbbk G$ is the free $\Bbbk$-module with basis $(b_g)_{g\in G}$, with product defined by $b_g b_{g'} := b_{gg'}$. The unit of the $\Bbbk$-algebra $\Bbbk G$ is the element $1 = b_e$ corresponding to the identity element $e \in G$.

The following exercise gives a relation between the homology of a group and the homology of its subgroups.

Exercise 2.31 (Shapiro's lemma). Let H be a subgroup of G. Let us denote by $\mathrm{res}^G_H : \Bbbk G\text{-Mod} \to \Bbbk H\text{-Mod}$ the restriction functor. If M is a $\Bbbk H$-module, we denote by $\mathrm{ind}^G_H M$ the $\Bbbk$-module $\Bbbk G \otimes_{\Bbbk H} M$, acted on by G by the formula: $g(x \otimes m) := gx \otimes m$. We denote by $\mathrm{coind}^G_H M$ the $\Bbbk$-module $\mathrm{Hom}_{\Bbbk H}(\Bbbk G, M)$, acted on by G by the formula $(gf)(x) := f(xg)$.

1. Prove that ind^G_H (resp. coind^G_H) is left (resp. right) adjoint to res^G_H.
2. Prove that res^G_H is exact and preserves injectives and projectives, that ind^G_H is exact and preserves projectives, and that coind^G_H is exact and preserves injectives (use Exercise 2.22).
3. Prove Schapiro's lemma, namely we have isomorphisms, natural with respect to the $\Bbbk H$-module N:
$$H_*(H, N) \simeq H_*(G, \mathrm{ind}^G_H N)\ , \quad H^*(H, N) \simeq H^*(G, \mathrm{coind}^G_H N)\ .$$

2.4.2. Products in cohomology. If A is a $\Bbbk G$-algebra, the cohomology $H^*(G, A)$ is equipped with a so-called cup product, which makes it into a graded $\Bbbk$-algebra. Moreover, if A is commutative, then $H^*(G, A)$ is graded commutative, i.e., the cup product of homogeneous elements x and y satisfies
$$x \cup y = (-1)^{\deg(x)\deg(y)} y \cup x\ .$$
This cup product may be defined in many different ways (there are four different definitions in [ML]). We present here a description of cup products using the complex $\mathrm{Hom}_{\Bbbk G}(B(\Bbbk G, \Bbbk G, \Bbbk^{\mathrm{triv}}), M)$ from Corollary 2.29. We first rewrite this complex under a more convenient form.

Proposition 2.32. *The complex* $\mathrm{Hom}_{\Bbbk G}(B(\Bbbk G, \Bbbk G, \Bbbk^{\mathrm{triv}}), M)$ *is isomorphic to the complex of* $\Bbbk$*-modules* $C^*(G, M)$ *defined by*
$$C^n(G, M) = \mathrm{Map}(G^{\times n}, M)\ ,$$
where $\mathrm{Map}(G^{\times n}, M)$ *means the maps from the set* $G^{\times n}$ *to the* $\Bbbk$*-module* M*, and* $G^{\times 0}$ *should be understood as a set with one element, and with differential* $\partial : C^n(G, M) \to C^{n+1}(G, M)$ *defined by:*
$$\begin{aligned}(\partial f)(g_1, \dots, g_{n+1}) := g_1 f(g_2, \dots, g_{n+1}) \\ + \textstyle\sum_{i=1}^n (-1)^i f(g_1, \dots, g_i g_{i+1}, \dots, g_{n+1}) \\ + (-1)^{n+1} f(g_2, \dots, g_n)\ .\end{aligned}$$

Proof. There is an isomorphism of $\Bbbk$-modules:
$$\mathrm{Map}(G^{\times n}, M) \simeq \mathrm{Hom}_{\Bbbk G}(\Bbbk G \otimes (\Bbbk G)^{\otimes n}, M)$$
which sends a map f to the $\Bbbk G$-linear morphism $\widetilde{f}$ defined by:
$$\widetilde{f}(b_{g_0}[b_{g_1}|\cdots|b_{g_n}]) := g_0 f(g_1, \dots, g_n)\ .$$
We check that this isomorphism is compatible with the differentials. $\square$

We define the cup product on the complex $C^*(G,A)$:

$$\begin{array}{ccc} C^n(G,A)\times C^m(G,A) & \to & C^{m+n}(G,A) \\ f_1,f_2 & \mapsto & f_1\cup f_2 \end{array}$$

by the formula (where $\cdot_A$ denotes the product of the $\Bbbk G$ algebra A)

$$(f_1\cup f_2)(g_1,\dots,g_{n+m}) := f_1(g_1,\dots,g_n)\cdot_A\left[(g_1\dots g_n)f_2(g_{n+1},\dots,g_{n+m})\right] .$$

This products makes $(C^*(G,A),\cup,\partial)$ into a differential graded algebra, that is, the differential ∂ of $C^*(G,A)$ acts on products by the formula:

$$\partial(f_1\cup f_2) = (\partial f_1)\cup f_2 + (-1)^{\deg(f_1)} f_1\cup(\partial f_2) .$$

Since $(C^*(G,A),\cup,\partial)$ is a differential graded algebra, there is an induced cup product on the level of cohomology, defined by the formula (where the brackets stand for the cohomology class represented by a cycle):

$$[f_1]\cup[f_2] := [f_1\cup f_2] .$$

Remark 2.33. If A is commutative, the differential graded algebra $C^*(G,A)$ is *not* graded commutative, so it is not clear with our definition why $H^*(G,A)$ should be graded commutative. One has to show that the difference $f_1\cup f_2 - (-1)^{\deg(f_1)\deg(f_2)} f_2\cup f_1$ is a boundary in $C^*(G,A)$, which is left as an exercise to the courageous reader.

2.5. Homology and cohomology of categories

As a reference for the cohomology of categories, the reader can consult [FP], or the article [Mit].

2.5.1. From groups to categories. A group G can be thought of as a category with one object, say $*$, with $\mathrm{Hom}(*,*) = G$ and the composite of morphisms g and h is the product gh. Let us denote by $*_G$ the category corresponding to the group G. With this description of a group, a $\Bbbk$-linear representation of G is the same as a functor

$$F : *_G \to \Bbbk\text{-Mod} .$$

The $\Bbbk$-module on which G acts is $F(*)$ and an element $g\in G$ acts on $F(*)$ by the $\Bbbk$-linear endomorphism $F(g)$. Keeping this in mind, the following definition is quite natural.

Definition 2.34. Let $\mathcal{C}$ be a small[4] category and let $\Bbbk$ be a commutative ring. A left $\Bbbk$-linear representation of $\mathcal{C}$ is a functor $F:\mathcal{C}\to\Bbbk$-Mod, and a right $\Bbbk$-linear representation is a functor $F:\mathcal{C}^{\mathrm{op}}\to\Bbbk$-Mod. A morphism of representations $\theta : F\to G$ is a natural transformation of functors.

Since the (co)homology of groups was defined in terms of Tor and Ext in the category of representations of G, the (co)homology of categories should be similarly defined in terms of Tor and Ext in the category of representations of $\mathcal{C}$. This motivates the study of functor categories.

[4] A *small* category is a category whose objects form a set. For example, the category of all sets is not a small category (cf. the classical paradox of the set of all sets).

2.5.2. The Yoneda lemma. The Yoneda lemma is an elementary result, but it plays a so crucial role in the study of categories and functors that it deserves its own section. Let $\mathcal{C}$ be a category and let x be an object of $\mathcal{C}$. We denote by $h^x : \mathcal{C} \to \mathrm{Set}$ the functor $c \mapsto \mathrm{Hom}_{\mathcal{C}}(x, c)$.

Lemma 2.35 (The Yoneda lemma). *Let $F : \mathcal{C} \to \mathrm{Set}$ be a functor, and let x be an object of $\mathcal{C}$. Then the natural transformations from h^x to F form a set, and the following map is a bijection:*

$$\begin{array}{ccc} \mathrm{Nat}(h^x, F) & \to & F(x) \\ \theta & \mapsto & \theta_x(\mathrm{Id}_x) \end{array} .$$

Proof. First, a natural transformation θ is uniquely determined by $\theta_x(\mathrm{Id}_x)$, since for all objects y in $\mathcal{C}$ and for all $f \in h^x(y)$

$$\theta_y(f) = (\theta_y \circ h^x(f))\,(\mathrm{Id}_x) = (F(f) \circ \theta_x)\,(\mathrm{Id}_x) = F(f)(\,\theta_x(\mathrm{Id}_x)\,)\,.$$

Hence the natural transformation forms an set and the map of Lemma 2.35 is injective. To prove it is surjective, we observe that if $\alpha \in F(x)$, the maps

$$\begin{array}{cccc} \theta^\alpha_y : & h^x(y) & \to & F(y) \\ & f & \mapsto & F(f)(\alpha) \end{array}$$

define a natural transformation θ^α in the preimage of α. □

Exercise 2.36. Let $f \in \mathrm{Hom}_{\mathcal{C}}(x, y)$. Show that the maps

$$\begin{array}{cccc} (h^f)_z : & h^y(z) & \to & h^x(z) \\ & \alpha & \mapsto & \alpha \circ f \end{array}$$

define a natural transformation $h^f : h^y \to h^x$. Let $\mathcal{H}$ be the category with the functors h^x as objects, and with natural transformations as morphisms. Show that the functor $\mathcal{C}^{\mathrm{op}} \to \mathcal{H}$, $x \mapsto h^x$ is an equivalence of categories.

2.5.3. The structure of functor categories. Let $\mathcal{C}$ be a small category and let $\Bbbk$ be a commutative ring. We denote by $\mathcal{C}$-Mod the category[5] whose objects are functors $F : \mathcal{C} \to \Bbbk$-Mod, and whose morphisms are natural transformations of functors.

Direct sums, products, kernels, cokernel, quotients of functors are defined in the target category. Specifically, this means that the direct sum of two functors F, G is the functor $F \oplus G$ defined by $(F \oplus G)(c) = F(c) \oplus G(c)$ for all $c \in \mathcal{C}$. Similarly, the kernel of a natural transformation $\theta : F \to G$ is defined by $(\ker \theta)(c) := \ker[\theta_c : F(c) \to G(c)]$ and so on. The following lemma is an easy check.

Lemma 2.37. *The category $\mathcal{C}$-Mod is an abelian category.*

[5] Actually we are cheating a bit here. It is a priori not clear that $\mathcal{C}$-Mod is a genuine category. Indeed, it is not clear that the collection of natural transformations between two functors form a set. However, this is the case when the source category $\mathcal{C}$ is small, and this is exactly why we impose this condition on $\mathcal{C}$.

Short exact sequences of functors are the diagrams of functors $0 \to F' \to F \to F'' \to 0$ such that for all $c \in \mathcal{C}$, evaluation on c yields a short exact sequence of $\Bbbk$-modules:

$$0 \to F'(c) \to F(c) \to F''(c) \to 0\,.$$

Since we consider functors with values in $\Bbbk$-mod, a linear combination of natural transformations is a natural transformation. So, the Hom-set in $\mathcal{C}$-Mod are actually $\Bbbk$-modules. Moreover, composition of natural transformations is bilinear. In other words, the functor category $\mathcal{C}$-Mod is enriched over $\Bbbk$ (we also say a $\Bbbk$-linear category).

2.5.4. Homological algebra in functor categories. We want to do homological algebra in $\mathcal{C}$-Mod. Let us first define the standard projectives P_x for $x \in \mathcal{C}$. These functors will play the same role in $\mathcal{C}$-Mod as the representation $\Bbbk G$ does in the category $\Bbbk G$-Mod.

Fix an object $x \in \mathcal{C}$. We define P_x as the $\Bbbk$-linearization of the functors h^x from Section 2.5.2. To be more specific, if X is a set, we let $\Bbbk X$ be the free $\Bbbk$-module with basis $(b_x)_{x\in X}$ indexed by the elements of X. Then $P_x \in \mathcal{C}$-Mod is defined by

$$P_x(c) := \Bbbk\,\mathrm{Hom}_{\mathcal{C}}(x,c)\,.$$

Lemma 2.38 ($\Bbbk$-linear Yoneda lemma). *Let $x \in C$ and let $F \in \mathcal{C}$-Mod. Let us denote by $1_x \in \Bbbk\mathrm{End}_{\mathcal{C}}(x)$ the basis element indexed by $\mathrm{Id}_x \in \mathrm{End}_{\mathcal{C}}(x)$. The following map is an isomorphism of $\Bbbk$-modules, natural with respect to F and x:*

$$\begin{array}{ccc}\mathrm{Hom}_{\mathcal{C}\text{-Mod}}(P_x,F) & \to & F(x)\\ \theta & \mapsto & \theta_x(1_x)\end{array}\,.$$

Proof. The canonical inclusion of sets $X \hookrightarrow \Bbbk\,X$ induces an isomorphism natural with respect to the set X and the $\Bbbk$-module M:

$$\mathrm{Hom}_{\Bbbk\text{-Mod}}(\Bbbk\,X, M) \simeq \mathrm{Hom}_{\mathrm{Set}}(X,M)\,.$$

Hence the canonical natural transformation of functors $h^x \to P^x$ induces an isomorphism $\mathrm{Hom}_{\mathcal{C}\text{-Mod}}(P_x,F) \simeq \mathrm{Nat}(h^x,F)$. Now the $\Bbbk$-linear map of Lemma 2.38 is the composite of this isomorphism and the standard Yoneda isomorphism of Lemma 2.35, hence it is bijective. □

Exercise 2.39. Let $\mathcal{C}$ be a small category, and let x,y be objects of $\mathcal{C}$. Write an explicit basis of $\mathrm{Hom}_{\mathcal{C}\text{-Mod}}(P_x,P_y)$.

Let us spell out the homological consequences of the Yoneda lemma.

Lemma 2.40. *Let $x \in \mathcal{C}$. The functor P_x is projective.*

Proof. Projectivity of P_x means exactness of the functor $\mathrm{Hom}_{\mathcal{C}\text{-Mod}}(P_x,-)$. By the Yoneda lemma, the latter is isomorphic to the functor $F \mapsto F(x)$, which is exact by the definition of short exact sequences of functors. □

Lemma 2.41. *The category $\mathcal{C}$-Mod has enough projectives, and the family of functors $(P_x)_{x\in\mathcal{C}}$ is a projective generator. That is, all $F \in \mathcal{C}$-Mod can be written as a quotient of a direct sum of these functors.*

Proof. By the Yoneda lemma, for all $x \in \mathcal{C}$ and all $\alpha \in F(x)$, there is a unique natural transformation $\widetilde{\alpha} : P_x \to F$ which sends $1_x \in P_x(x)$ to $\alpha \in F(x)$. Taking the sum of all the $\widetilde{\alpha}$ we obtain a surjective natural transformation:

$$\bigoplus_{x\in\mathcal{C},\alpha\in F(x)} P_x \twoheadrightarrow F. \qquad \square$$

Exercise 2.42. Prove that any projective object in $\mathcal{C}$-Mod can be seen as a direct summand of a direct sum of P_x.

The category $\mathcal{C}$-Mod also has enough injectives, but it is slightly more complicated to prove.

Exercise 2.43. Let J be an injective generator of $\Bbbk$-Mod (i.e., every $\Bbbk$-module embeds into a product of copies of J). For all $x \in \mathcal{C}$, let us denote by $I_{x,J} \in \mathcal{C}$-Mod the functor defined by:

$$I_{x,J}(c) := \mathrm{Hom}_{\Bbbk}\left(\Bbbk\,\mathrm{Hom}_{\mathcal{C}}(c,x), J\right) .$$

Show that $I_{x,J}$ is injective, and that the family $(I_{x,J})_{x\in\mathcal{C}}$ is an injective generator of $\mathcal{C}$-Mod. (Find inspiration from the proof of Proposition 2.12.)

2.5.5. Homology and cohomology of categories. Let $\mathcal{C}$ be a small category, let $\Bbbk$ be a commutative ring and let $F : \mathcal{C} \to \Bbbk$-Mod be a $\Bbbk$-linear representation of $\mathcal{C}$. Let us denote by $\Bbbk$ the constant functor with value $\Bbbk$. The cohomology of $\mathcal{C}$ with coefficients in F is the graded $\Bbbk$-module $H^*(\mathcal{C}, F)$ defined by

$$H^*(\mathcal{C}, F) := \mathrm{Ext}^*_{\mathcal{C}\text{-Mod}}(\Bbbk, F) .$$

To define the homology $H_*(\mathcal{C}, F)$ of $\mathcal{C}$ with coefficients in the representation F, we need a generalization of tensor products to the framework of functors. Let us denote the category $\mathcal{C}^{\mathrm{op}}$-Mod of contravariant functors with source $\mathcal{C}$ and target $\Bbbk$-Mod by the more suggestive notation Mod-$\mathcal{C}$. Such contravariant functors will play the role of right modules in our tensor product definition.

Let $G \in$ Mod-$\mathcal{C}$ and let $F \in \mathcal{C}$-Mod. The tensor product $G \otimes_{\mathcal{C}} F$ is the $\Bbbk$-module generated by the symbols $m\otimes n$, for $m \in G(x)$ and $n \in F(x)$, for all $x \in \mathcal{C}$, subject to the relations:

$$m \otimes n + m' \otimes n = (m + m') \otimes n ,$$
$$m \otimes n + m \otimes n' = (m + m') \otimes n ,$$
$$\lambda(m \otimes n) = (m\lambda) \otimes n = m \otimes (\lambda n) \text{ for } \lambda \in \Bbbk ,$$
$$(G(f)(m)) \otimes n = m \otimes (F(f)(n)) \text{ for all morphisms } f \text{ in } \mathcal{C} .$$

Exercise 2.44. Prove that $P_x \otimes_{\mathcal{C}} F \simeq F(x)$ naturally with respect to F, x and similarly that $G \otimes_{\mathcal{C}} P_x \simeq G(x)$, naturally with respect to G, x.

Exercise 2.45. Prove that the isomorphisms, natural with respect to $G \in \text{Mod-}\mathcal{C}$, $F \in \mathcal{C}\text{-Mod}$ and $M \in \Bbbk\text{-Mod}$:

$$\begin{aligned}\mathrm{Hom}_{\Bbbk\text{-Mod}}(G \otimes_{\mathcal{C}} F, M) &\simeq \mathrm{Hom}_{\text{Mod-}\mathcal{C}}(G, \mathrm{Hom}_{\Bbbk}(F, M))\;,\\ \mathrm{Hom}_{\Bbbk\text{-Mod}}(G \otimes_{\mathcal{C}} F, M) &\simeq \mathrm{Hom}_{\mathcal{C}\text{-Mod}}(F, \mathrm{Hom}_{\Bbbk}(G, M))\;.\end{aligned}$$

Deduce from these isomorphisms the following properties of the functors $F\otimes_{\mathcal{C}}$ and $\otimes_{\mathcal{C}}F$: (i) they are right exact, (ii) they commute with arbitrary sums, and (iii) if F (resp. G) is projective, then $G\otimes_{\mathcal{C}}$ (resp. $\otimes_{\mathcal{C}}F$) is exact.

As in the case of modules over an algebra, we define the Tor functors by deriving the right exact functors $G\otimes_{\mathcal{C}} : \mathcal{C}\text{-Mod} \to \Bbbk\text{-Mod}$ and $\otimes_{\mathcal{C}}F : \text{Mod-}\mathcal{C} \to \Bbbk\text{-Mod}$. As in the case of modules over a ring, the notation has two interpretations, but we can prove that they coincide. Thus we have:

$$\mathrm{Tor}_i^{\mathcal{C}}(G, F) = L_i(G\otimes_{\mathcal{C}})(F) = L_i(\otimes_{\mathcal{C}}F)(G)\;.$$

The homology $H_*(\mathcal{C}, F)$ of a small category $\mathcal{C}$ with coefficients in the functor $F \in \mathcal{C}\text{-Mod}$ is then defined by:

$$H_*(\mathcal{C}, F) := \mathrm{Tor}_*^{\mathcal{C}}(\Bbbk, F)\;.$$

2.6. Cohomology of linear algebraic groups

We present three equivalent viewpoints on linear algebraic groups and their representations. We may view a linear algebraic groups as a Zariski closed subgroup of matrices (this viewpoint requires that the ground field $\Bbbk$ is algebraically closed, and the reader may take [Bo], [Hum] and [S] as references), as a commutative finitely generated Hopf $\Bbbk$-algebra, or as a representable functor from finitely generated $\Bbbk$-algebras to groups (we refer to [Wa] for the latter viewpoints, which are valid over an arbitrary ground field). Then we define their cohomology in terms of Exts. We recall the definition of the Hochschild complex, and we use it to define the cohomological product. References for the cohomology of algebraic groups are [F] and [J].

2.6.1. Linear algebraic groups and their representations. Let $\Bbbk$ be an algebraically closed field. A *linear algebraic group over* $\Bbbk$ is a Zariski closed subgroup of some $SL_n(\Bbbk)$, i.e., which is defined as the zero set of a family of polynomials with coefficients in $\Bbbk$ and with variables in the entries of the matrices $[x_{i,j}] \in SL_n(\Bbbk)$.

Example 2.46. Besides finite groups, we have the following common examples of algebraic groups:

$$\begin{aligned}SL_n(\Bbbk) &= \{M \in M_n(\Bbbk)\ ;\ \det M = 1\}\;,\\ GL_n(\Bbbk) &= \left\{\begin{bmatrix} * & 0\\ 0 & M\end{bmatrix} \in SL_{n+1}(\Bbbk)\right\}\;,\\ \mathbb{G}_a &= \left\{\begin{bmatrix} 1 & *\\ 0 & 1\end{bmatrix} \in M_2(\Bbbk)\right\}\;,\\ O_n(\Bbbk) &= \{M \in M_n(\Bbbk)\ ;\ M^tM = I\}\;,\\ Sp_n(\Bbbk) &= \{M \in M_{2n}(\Bbbk)\ ;\ M^t\Omega M = I\}\;,\end{aligned}$$

where Ω is the matrix in the canonical basis (e_i) of $\Bbbk^{2n}$ of the bilinear form ω_n defined by $\omega_n(e_i, e_j) = 1$ if $j = i + n$, $\omega_n(e_i, e_j) = -1$ if $i + n = j$, and $\omega_n(e_i, e_j)$ is zero otherwise.

A *rational (or algebraic) representation* of a linear algebraic group G is a $\Bbbk$-vector space V, equipped with an action $\rho : G \to \mathrm{End}_{\Bbbk}(V)$, satisfying the following condition.

(C) If V is finite-dimensional, choose a basis of V and let $\rho_{k,\ell}(g)$ be the coordinates of $\rho(g)$ in the corresponding basis of $\mathrm{End}_{\Bbbk}(V)$. Then we require that the maps $g \mapsto \rho_{k,\ell}(g)$ are polynomials in the entries of the matrices $[g_{ij}] \in G$. (One says that ρ is *regular*).

(C') If V is infinite-dimensional, we require that V is the union of finite-dimensional sub-representations satisfying (C).

Example 2.47.

1. If V is a $\Bbbk$-vector space, the representation V^{triv} (i.e., G acts trivially) is a rational representation.
2. If $G \subset SL_n(\Bbbk)$, then $\Bbbk^n$ acted on by G by matrix multiplication is a rational representation.
3. If $G \subset SL_n(\Bbbk)$ is an algebraic group, let $\Bbbk[G]$ be the regular functions on G. That is, $\Bbbk[G]$ is the quotients of the algebra $\Bbbk[x_{i,j}]_{1 \le i,j \le n}$ by the ideal of the polynomials vanishing on G. The action of G on $\Bbbk[G]$ by left translations $(gf)(x) := f(g^{-1}x)$ is a rational action.
4. If V and W are rational representations of G, the diagonal action on $V \otimes_{\Bbbk} W$ (that is $g(v \otimes w) := (gv) \otimes (gw)$ makes it into a rational representation of G.

We denote by rat-G-Mod the category whose objects are rational representations of G and morphisms are G-equivariant maps. Thus, rat-G-Mod is a full subcategory of $\Bbbk G$-mod. It is not hard to check that direct sums, sub-representations and quotients of rational representations are rational representations, so rat-G-Mod is actually a full abelian subcategory of $\Bbbk G$-Mod. The following exercise provides an explanation for the name 'rational G-module'.

Exercise 2.48. Show that a morphism of monoids $\rho : GL_n(\Bbbk) \to M_m(\Bbbk)$ defines a rational representation of $GL_n(\Bbbk)$ if and only if its coordinates $\rho_{k,\ell}$ satisfy the following property. There exists a polynomial $P_{k,\ell}(x_{i,j})$ with n^2 variables $(1 \le i, j \le n)$ and an integer d such that for all $[g_{i,j}]_{1 \le i,j \le n} \in GL_n(\Bbbk)$ we have:

$$\rho_{k,\ell}([g_{i,j}]) = \frac{P(g_{i,j})}{(\det[g_{i,j}])^d} \, .$$

2.6.2. From algebraic groups to Hopf algebras. The definitions of Section 2.6.1 are not relevant over general fields $\Bbbk$. For example, if $\Bbbk$ is a finite field, then all subsets of $M_n(\Bbbk)$ are Zariski closed, and all maps $M_n(\Bbbk) \to M_m(\Bbbk)$ may be expressed as values of polynomials. Thus, representations of algebraic groups would have the same meaning as representations of finite groups. This is certainly not what we want.

To bypass this problem, we use the coordinate algebra $\Bbbk[G]$ of a linear algebraic group $G \subset M_n(\Bbbk)$. It is the finitely generated commutative reduced (i.e., without nilpotents) $\Bbbk$-algebra obtained as the quotient of $\Bbbk[x_{i,j}, 1 \leq i,j \leq n]$ by the ideal of polynomials vanishing on G. It may be interpreted a the algebra of regular maps $G \to \Bbbk$, that is set-theoretic maps which can be obtained as restrictions of polynomials with n^2 variables. The $\Bbbk$-algebra $\Bbbk[G]$ characterizes G as a Zariski closed set but says nothing about the group structure of G.

Proposition 2.49. *Let G be a linear algebraic group over an algebraically closed field $\Bbbk$. The group structure of G yields a Hopf algebra structure on $\Bbbk[G]$.*

Recall that a *Hopf $\Bbbk$-algebra* H is a $\Bbbk$-algebra H equipped with morphisms of algebras (respectively called the comultiplication, the counit and the antipode):

$$H \xrightarrow{\Delta} H^{\otimes 2}\;, \quad H \xrightarrow{\epsilon} \Bbbk\;, \quad H \xrightarrow{\chi} H\;,$$

which satisfy the following axioms (where all tensor products are taken over $\Bbbk$ and $\eta : \Bbbk \to H$ is the morphism $\lambda \mapsto \lambda 1$):

(i) Δ is coassociative, i.e.: $(\Delta \otimes \mathrm{Id}_H) \circ \Delta = (\mathrm{Id}_H \otimes \Delta) \circ \Delta$,
(ii) Δ is counital, i.e.: $(\mathrm{Id}_H \otimes \epsilon) \circ \Delta = \eta \circ \epsilon = (\epsilon \otimes \mathrm{Id}_H) \circ \Delta$,
(iii) Δ has a coinverse, i.e.: $m \circ (\chi \otimes \mathrm{Id}_H) \circ \Delta = \mathrm{Id}_H = m \circ (\mathrm{Id}_H \otimes \chi) \circ \Delta$.

A *morphism of Hopf algebras* $f : H \to H'$ is a $\Bbbk$-algebra morphism which commutes with comultiplications, counits and antipodes:

$$(f \otimes f) \circ \Delta = \Delta' \circ f\;, \quad \epsilon' \circ f = \epsilon\;, \quad \chi' \circ f = f \circ \chi\;.$$

Proof of Proposition 2.49. We respectively define Δ, χ and ϵ as precomposition of regular functions on G by the multiplication, the inverse and the unit of G. The axioms (i), (ii), (iii) follow directly from the associativity of m, the fact that m has a unit, and the equation $g \cdot g^{-1} = 1 = g^{-1} \cdot g$. □

Example 2.50. Let $\Bbbk$ be an algebraically closed field. The $\Bbbk$-algebra of regular functions of the algebraic group $GL_n(\Bbbk)$ is

$$\Bbbk[x_{i,j}, t] / \langle t \det[x_{i,j}] - 1 \rangle\;,$$

where there are n^2 variables $x_{i,j}$ $(1 \leq i,j \leq n)$, and the brackets on the right refer to the ideal generated by the polynomial $t \det[x_{i,j}] - 1$. The comultiplication, the counit and the antipode are given by:

$$\begin{aligned} \Delta(x_{i,j}) &= \sum_{k=1}^{n} x_{i,k} \otimes x_{k,j}\;, & \Delta(t) &= t \otimes t\;, \\ \epsilon(x_{i,j}) &= \delta_{i,j}\;, & \epsilon(t) &= 1\;, \\ \chi(x_{i,j}) &= t(-1)^{ij}\mathrm{M}_{i,j} & \chi(t) &= \det[x_{i,j}]\;, \end{aligned}$$

where $\mathrm{M}_{i,j}$ refers to the polynomial of degree $(n-1)^2$ obtained as the (i,j)-minor of the $n \times n$ matrix $[x_{k,\ell}]_{1 \leq k,\ell \leq n}$.

Rational representations of linear algebraic groups can be translated in the language of Hopf algebras.

Proposition 2.51. *Let $\Bbbk$ be an algebraically closed field, and let G be an algebraic group scheme over $\Bbbk$. The category of rational G-modules is equivalent to the category Comod-$\Bbbk[G]$ of right $\Bbbk[G]$-comodules.*

Recall that if $(H, \Delta, \epsilon, \chi)$ is a Hopf algebra over $\Bbbk$, a *right H-comodule* is a $\Bbbk$-vector space V equipped with a coaction morphism $\Delta_V : V \to V \otimes_{\Bbbk} H$ satisfying the following axioms:

(i) compatibility with Δ, i.e.: $(\mathrm{Id}_V \otimes \Delta) \circ \Delta_V = (\Delta_V \otimes \mathrm{Id}_H) \circ \Delta_V$,
(ii) compatibility with ϵ, i.e.: $(\mathrm{Id}_V \otimes \epsilon) \circ \Delta_V = \mathrm{Id}_V$.

A *morphism of comodules* is a $\Bbbk$-linear morphism $f : V \to W$ which commutes with the coaction, i.e., $(f \otimes \mathrm{Id}_H) \circ \Delta_V = \Delta_W \circ f$. If H is a Hopf algebra over a field $\Bbbk$, the category Comod-H of right H-comodules is abelian.

Proof of Proposition 2.51. We first prove that the category of finite-dimensional rational representations of G is equivalent to the category of finite-dimensional $\Bbbk[G]$-comodules. We let $(b_i)_{0 \leq i \leq n}$ be the canonical basis of $\Bbbk^n$.

1. Let $\rho : G \to M_n(\Bbbk)$ be a rational representation, and let $\rho_{k,\ell}$, $1 \leq k, \ell \leq n$ denote its coordinates. We check that the following formula defines a $\Bbbk[G]$-comodule map on $\Bbbk^n$:

$$\Delta_\rho(b_\ell) = \textstyle\sum_{i=1}^n b_k \otimes \rho_{k,\ell} \,.$$

Moreover, we check that $\Bbbk$-linear map $f : \Bbbk^n \to \Bbbk^m$ is G equivariant if and only if it is a comodule map for the corresponding coactions. So, by sending a G-module ρ to the comodule given by Δ_ρ (and by acting identically on Hom-sets) we get a fully faithful functor:

$$\Phi : \{\text{fin. dim. rat-}G\text{-Mod}\} \to \{\text{fin. dim. } \Bbbk[G]\text{-comodules}\} \,.$$

2. Conversely, if $\Delta : \Bbbk^n \to \Bbbk^n \otimes \Bbbk[G]$ is a comodule structure on $\Bbbk^n$, then $\Delta_{\Bbbk^n}(b_j)$ can be uniquely written as a sum $\sum_{i=1}^n b_i \otimes \Delta_{i,j}$ with $\Delta_{i,j} \in \Bbbk[G]$. We check that the map $\rho : G \to M_n(\Bbbk)$ defined by $\rho(g) = [\Delta_{i,j}(g)]$ is a morphism of groups. Thus Φ is essentially surjective, hence it is an equivalence of categories.

Now we prove that rat-G-mod is equivalent to Comod-$\Bbbk[G]$.

3. Assume that V is the union of finite-dimensional vector subspaces $(V_\alpha)_{\alpha \in A}$, and that there are comodule maps $\Delta_\alpha : V_\alpha \to V_\alpha \otimes \Bbbk[G]$ such that the inclusions $V_\alpha \subset V_\beta$ are morphisms of comodules. Then we define a comodule structure $V \to V \otimes \Bbbk[G]$ by letting $\Delta(v) = \Delta_\alpha(v)$ for any α such that $v \in V_\alpha$ ($\Delta(v)$ does not depend on α).

In particular, the functor Φ extends to a fully faithful functor:

$$\overline{\Phi} : \text{rat-}G\text{-Mod} \to \text{Comod-}\Bbbk[G] \,.$$

4. To prove that $\overline{\Phi}$ is an equivalence of categories, it remains to prove that it is essentially surjective, which is equivalent to prove that all $\Bbbk[G]$-comodules can be obtained as the union of their finite-dimensional subcomodules. This finiteness property actually holds for comodules over arbitrary finitely generated Hopf algebras, see [Wa, Chap. 3.3].

This concludes the proof. □

To sum up, a linear algebraic group G over an algebraically closed field $\Bbbk$ yields a Hopf algebra $\Bbbk[G]$, and the study of the rational representations of G is equivalent to the study of the right $\Bbbk[G]$-comodules. Now when $\Bbbk$ is an arbitrary field, the definitions of Section 2.6.1 are not relevant, but their Hopf algebra version still is. So we just think of algebraic groups as commutative finitely generated Hopf $\Bbbk$-algebra, and of rational modules as right comodules.

Example 2.52. If $\Bbbk$ is an arbitrary field, the formulas of Example 2.50 define commutative finitely generated Hopf $\Bbbk$-algebras, which we denote $\Bbbk[GL_n]$. The category of rational representations of the general linear group over $\Bbbk$ is defined as the category of right $\Bbbk[GL_n]$-comodules.

Remark 2.53. If the ground field $\Bbbk$ is algebraically closed, the coordinate algebra $\Bbbk[G]$ of a linear algebraic group G is reduced, i.e., without nilpotent elements. So by considering *all* commutative finitely generated Hopf $\Bbbk$-algebras H, we introduce new objects. For example, if $\Bbbk$ has characteristic $p > 0$, the finite-dimensional Hopf algebra $\Bbbk[(GL_n)_r]$ obtained as a quotient of $\Bbbk[GL_n]$ by the relations $x_{i,j}^{p^r} = \delta_{i,j}$ has nilpotent elements. Such Hopf algebras (called Frobenius kernels of GL_n in the group scheme terminology) play a central role in the study of the rational representations of the general linear group, even when the ground field $\Bbbk$ is algebraically closed, see [J].

2.6.3. From Hopf algebras to affine group schemes. Let $\Bbbk$ be a field. An affine group scheme over $\Bbbk$ is a representable functor

$$G : \{\text{fin. gen. com. } \Bbbk\text{-Alg}\} \to \{\text{Groups}\} .$$

Representable means that there exists a finitely generated commutative $\Bbbk$-algebra B such that G is the functor $\mathrm{Hom}_{\Bbbk\text{-Alg}}(B, -)$. The algebra B representing G is often denoted by $\Bbbk[G]$. A morphism of affine group schemes is a natural transformation.

If H is a Hopf algebra we endow the set $\mathrm{Hom}_{\Bbbk\text{-Alg}}(H, A)$ with a group structure by letting the product of f and g be $m_A \circ (f \otimes g) \circ \Delta$, where m_A denotes the multiplication of A. (The unit element is the map $1_A\epsilon$, and the inverse of f is $f \circ \chi$). This group structure is natural with respect to A, hence the functor $\mathrm{Sp}(H) = \mathrm{Hom}_{\Bbbk\text{-Alg}}(H, -)$ has values in groups. This yields a functor

$$\mathrm{Sp} : \{\text{fin. gen. com. Hopf } \Bbbk\text{-alg}\}^{\mathrm{op}} \to \{\text{Affine Group Schemes over } \Bbbk\} .$$

As a consequence of the Yoneda lemma 2.35, this functor is an equivalence of categories (compare with Exercise 2.36). Thus affine group schemes are a geometric way to view Hopf algebras.

Exercise 2.54. Show that $\mathrm{Sp}(\Bbbk[GL_n])$ is the functor which sends a $\Bbbk$-algebra A to the group $GL_n(A)$.

Now we translate comodules over Hopf algebras in the language of affine group schemes. If V is a $\Bbbk$-vector space, we denote by End_V the functor from finitely generated $\Bbbk$-algebras to monoids, which sends an algebra A to $\mathrm{End}_A(V \otimes_\Bbbk A)$, where $V \otimes_\Bbbk A$ is considered as a A-module by the formula $a \cdot (v \otimes b) := v \otimes (ab)$. A representation of a group scheme G is simply a natural transformation of functors $\rho : G \to \mathrm{End}_V$. A $\Bbbk$-linear morphism $f : V \to W$ is a morphism of representations if for all A, the A-linear map $f \otimes \mathrm{Id}_A$ is $G(A)$-equivariant. We denote by G-Mod the abelian category of representations of G. Let H be a finitely generated commutative Hopf $\Bbbk$-algebra. Using the Yoneda lemma once again, we prove an equivalence of categories:

$$\text{Comod-}H \simeq \mathrm{Sp}(H)\text{-Mod}\ .$$

2.6.4. Cohomology of algebraic groups or group schemes. We have several ways to think of representations of linear algebraic groups or group schemes. For example we have three equivalent categories for the general linear group: the representations of the group scheme GL_n, the comodules over the Hopf algebra $\Bbbk[GL_n]$, and the rational representations of the linear algebraic group $GL_n(\Bbbk)$ (the latter is well defined over an algebraically closed field only)

$$GL_n\text{-Mod} \simeq \text{Comod-}\Bbbk[GL_n] \simeq \text{rat-}GL_n(\Bbbk)\text{-mod}\ .$$

The reader may work with his favourite category. From now on, we fix an affine group scheme G and we focus on the category G-Mod in the statements, although we might use another version of this category in the proofs.

Proposition 2.55. *The abelian category* G-Mod *has enough injectives.*

Proof. To prove the proposition, it suffices to prove that all $\Bbbk[G]$-comodule V can be embedded into an injective $\Bbbk[G]$-comodule.

1. Let us denote by $V^{\mathrm{triv}} \otimes \Bbbk[G]$ the $\Bbbk$-vector space $V \otimes \Bbbk[G]$ equipped with the coaction $\mathrm{Id}_V \otimes \Delta$. The assignment $f \mapsto (f \otimes \mathrm{Id}_{\Bbbk[G]}) \circ \Delta_W$ defines an isomorphism, natural with respect to $V \in \Bbbk$-mod and $W \in$ Comod-$\Bbbk[G]$:
$$\mathrm{Hom}_\Bbbk(W, V) \simeq \mathrm{Hom}_{\text{Comod-}\Bbbk[G]}(W, V^{\mathrm{triv}} \otimes \Bbbk[G])\ .$$
Since $\Bbbk$ is a field, the left-hand side is an exact functor with variable W. Whence the injectivity of $V^{\mathrm{triv}} \otimes \Bbbk[G]$.
2. For all $\Bbbk[G]$-comodules V, the $\Bbbk$-linear map $\Delta_V : V \to V^{\mathrm{triv}} \otimes \Bbbk[G]$ is injective (by the compatibility with ϵ) and it is a morphism of comodules (by the compatibility with Δ).

Thus V embeds in the injective comodule $V^{\mathrm{triv}} \otimes \Bbbk[G]$. □

As in the case of discrete groups, there is a fixed point functor

$$\begin{array}{rccl} -^G : & G\text{-Mod} & \to & \Bbbk\text{-vect} \\ & V & \mapsto & V^G = \mathrm{Hom}_{G\text{-Mod}}(\Bbbk^{\mathrm{triv}}, V) \end{array}$$

Since G-Mod has enough injectives, the right derived functors of the fixed point functor are well defined and we call them the rational cohomology of G (or simply the cohomology of G).

$$H^i(G, -) = R^i(-^G) = \mathrm{Ext}^i_{G\text{-Mod}}(\Bbbk^{\mathrm{triv}}, -)\ .$$

Remark 2.56. The category G-Mod has usually not enough projectives [J, I, Section 3.18]. So in general we cannot hope to derive a functor of coinvariants to define a notion of rational homology of an algebraic group.

If G is a linear algebraic group over an algebraically closed field $\Bbbk$ (as in Section 2.6.1), the notation $H^*(G, V)$ is ambiguous. Indeed, this notation might stand for its rational cohomology $\mathrm{Ext}^*_{\text{rat-}G\text{-Mod}}(\Bbbk^{\mathrm{triv}}, V)$, or for the cohomology of the discrete group G in the sense of Section 2.4, i.e., $\mathrm{Ext}^*_{\Bbbk G\text{-Mod}}(\Bbbk^{\mathrm{triv}}, V)$. These Exts need not be the same, since they are not computed in the same categories: rat-G-Mod is only a subcategory of $\Bbbk\, G$-Mod. It is usually clear from the context which notion is used. The following exercise should warn the reader of the dangers of misinterpreting the notation $H^*(G, V)$.

Exercise 2.57. Let $\Bbbk$ be an algebraically closed field.

1. Let $\rho : GL_1(\Bbbk) \to GL_n(\Bbbk)$ be a representation of the multiplicative group. Prove that there exists a family of $n \times n$-matrices $(M_k)_{k \in \mathbb{Z}}$ which are all zero but a finite number of them such that $\rho(x) = \sum x^k M_k$. Show that the M_k satisfy the relations

$$\sum M_k = I\ , \quad M_k M_\ell = 0 \text{ if } k \neq \ell\ , \quad M_k^2 = M_k\ .$$

Deduce that the representation splits as a direct sum of simple representations of dimension 1.

2. Prove that for all rational representations V of $GL_1(\Bbbk)$,

$$\mathrm{Ext}^*_{\text{rat-}GL_1(\Bbbk)\text{-Mod}}(\Bbbk^{\mathrm{triv}}, V) = 0 \text{ if } i > 0.$$

3. If $\Bbbk = \mathbb{R}$ or $\mathbb{C}$, prove that the following representation of the discrete group $GL_1(\Bbbk)$ provides a non-split extension of $\Bbbk^{\mathrm{triv}}$ by $\Bbbk^{\mathrm{triv}}$:

$$\begin{array}{rccc} \rho : & GL_1(\Bbbk) & \to & \mathrm{End}(\Bbbk^2) \\ & z & \mapsto & \begin{bmatrix} 1 & \ln(|z|) \\ 0 & 1 \end{bmatrix} \end{array}.$$

Deduce that $\mathrm{Ext}^1_{\Bbbk\, GL_1(\Bbbk)\text{-Mod}}(\Bbbk^{\mathrm{triv}}, \Bbbk^{\mathrm{triv}}) \neq 0$.

2.6.5. Hochschild complex and cohomology algebras. The Hochschild complex computes the rational cohomology $H^*(G, V)$ of an affine group scheme G with coefficients in V. This complex does not require specific knowledge about G or V to be written down (but of course, some knowledge is needed if you want to compute its homology!). It is often used to describe the general (i.e., independent of G and V) properties of rational cohomology. As an example, we will use it to define cohomology cup products.

The Hochschild complex is similar to the complex of Proposition 2.32 (computing the cohomology of discrete groups), but set-theoretic maps $G^{\times n} \to V$ have to be replaced by their algebro-geometric counterpart. To be more specific, if V is a finite-dimensional vector space, we can consider it as a scheme, that is a representable functor

$$\{\text{fin. gen. com. } \Bbbk\text{-Alg}\} \to \{\text{Sets}\} .$$

Its value on A is the set $V \otimes_{\Bbbk} A$, and it is represented by $\Bbbk[V]$. We shall write $\mathrm{Hom}(G^{\times n}, V)$ for the set of natural transformations from $G^{\times n}$ to V. If V is infinite-dimensional, we let $\mathrm{Hom}(G^{\times n}, V)$ be the set of natural transformations taking values in a finite-dimensional subspace of V. By the Yoneda lemma, we have a $\Bbbk$-linear isomorphism:

$$V \otimes \Bbbk[G]^{\otimes n} \simeq \mathrm{Hom}(G^{\times n}, V) .$$

We can now define the Hochschild complex (compare with Proposition 2.32). It is the complex of $\Bbbk$-modules $C^*(G, V)$ with:

$$C^n(G, V) = \mathrm{Hom}(G^{\times n}, V) \simeq V \otimes (\Bbbk[G]^{\otimes n}) ,$$

and the differential $\partial : C^n(G, V) \to C^{n+1}(G, V)$ sends a morphism f to the morphism ∂f defined by

$$\begin{aligned}(\partial f)(g_1, \dots, g_{n+1}) := & \; g_1 f(g_2, \dots, g_{n+1}) \\ & + \textstyle\sum_{i=1}^{n} (-1)^i f(g_1, \dots, g_i g_{i+1}, \dots, g_{n+1}) \\ & + (-1)^{n+1} f(g_2, \dots, g_n) .\end{aligned}$$

The proof of the following proposition is dual to the proof of Proposition 2.27 and Corollary 2.28, see [J, I Section 4.16]:

Proposition 2.58. *The homology of $C^*(G, V)$ is equal to $H^*(G, V)$.*

Let G be a group scheme and let A be a $\Bbbk G$-algebra. By this we mean that A is a $\Bbbk$-algebra, equipped with an action of G by algebra automorphisms. Then the rational cohomology of G with coefficients in A is equipped with a cup product. This cup product can be defined on the level of the Hochschild complex exactly as in the case of discrete groups. Namely, if $\cdot_A$ denotes the multiplication in the algebra A, we define the cup product of $f_1 \in C^n(G, A)$ and $f_2 \in C^m(G, A)$ by:

$$(f_1 \cup f_2)(g_1, \dots, g_{n+m}) := f_1(g_1, \dots, g_n) \cdot_A [(g_1 \cdots g_n) f_2(g_{n+1}, \dots, g_{n+m})] .$$

This product makes the Hochschild complex into a differential graded algebra, hence it induces a product on homology. As in the case of discrete groups, one can prove that if A is commutative, then the graded algebra $H^*(G, A)$ is graded commutative.

3. Derived functors of non-additive functors

Many functors between abelian categories are not additive. In this section, we explain how to derive arbitrary functors between abelian categories. This theory was invented by Dold and Puppe, and later generalized by Quillen's homotopical algebra. Dold–Puppe theory of derived functors relies heavily on simplicial techniques, for which the reader can take [GJ, May, ML, Wei] as references. The best reference for derived functors is the seminal article [DP].

3.1. Simplicial objects

3.1.1. The categorical definition. Let Δ denote the category whose objects are the finite-ordered sets $[n] = \{0 < 1 < \cdots < n\}$ for $n \geq 0$, and whose morphisms are the non-decreasing monotone maps.

Definition 3.1. Let $\mathcal{C}$ is a category, a *simplicial object in* $\mathcal{C}$ is a functor $\Delta^{\mathrm{op}} \to \mathcal{C}$. A *morphism of simplicial objects* is a natural transformation. We denote by $s\mathcal{C}$ the category of simplicial objects in $\mathcal{C}$.

As a functor category, $s\mathcal{C}$ inherits the properties of $\mathcal{C}$. For example, if $\mathcal{C}$ has products then so does $s\mathcal{C}$, the product $X \times Y$ is defined by $(X \times Y)([n]) := X([n]) \times Y([n])$. If $\mathcal{C}$ is an abelian category, so is $s\mathcal{C}$. If $F : \mathcal{C} \to \mathcal{D}$ is a functor, postcomposition by F induces a functor $s\mathcal{C} \to s\mathcal{D}$.

3.1.2. The geometric definition. The categorical definition of simplicial objects is practical to derive some of their abstract properties, but there is an equivalent geometrical definition which gives a more concrete picture of what simplicial objects are.

For $0 \leq i \leq n$ we denote by $d^i : [n-1] \to [n]$ the unique injective morphism of Δ which misses $i \in [n]$, and by $s^i : [n+1] \to [n]$ the unique surjective map in Δ with two elements sent to $i \in [n]$. It can be proved that the d^i and the s^i generate the category Δ, that is all morphism of Δ can be written as a composite of these maps. Thus, a simplicial object X in $\mathcal{C}$ is uniquely determined by the sequence of objects $X_n = X([n])$, $n \geq 0$, and by the morphisms

$$d_i := X(d^i) : X([n]) \to X([n-1])\,, \quad s_i := X(s^i) : X([n]) \to X([n+1])\,.$$

However, the category Δ is not freely generated by the d^i and the s^i: relations hold between various compositions of these maps. So not all data of objects $(X_n)_{n\geq 0}$ and morphisms $d_i : X_n \to X_{n-1}$ and $s_i : X_n \to X_{n+1}$ come from simplicial objects. One can prove that the ones corresponding to simplicial objects are the ones satisfying the following simplicial identities.

$$\begin{aligned}
d_i d_j &= d_{j-1} d_i \text{ if } i < j,\\
s_i s_j &= s_{j+1} s_i \text{ if } i \leq j,\\
d_i s_j &= \begin{cases} s_{j-1} d_i & \text{if } i < j,\\ \mathrm{id} & \text{if } i = j \text{ or } i = j+1,\\ s_j d_{j-1} & \text{if } i > j+1. \end{cases}
\end{aligned}$$

This leads us to the geometric definition of simplicial objects.

Definition 3.2. Let $\mathcal{C}$ be a category. A *simplicial object in* $\mathcal{C}$ is a sequence of objects $(X_n)_{n\geq 0}$ of $\mathcal{C}$, together with face operators $d_i : X_n \to X_{n-1}$ and degeneracy operators $s_i : X_n \to X_{n+1}$ for $0 \leq i \leq n$, satisfying the simplicial identities listed above. A *morphism of simplicial sets* is a sequence of maps $f_n : X_n \to Y_n$, $n \geq 0$ commuting with the face and the degeneracy operators.

Of most interest for us are the category sSet of simplicial sets and the sR-Mod of simplicial R-modules.

3.1.3. Examples of simplicial sets. The *standard n-simplex* $\Delta[n]$ is the simplicial set defined by:

$$\Delta[n] := \mathrm{Hom}_\Delta(-, [n]) .$$

The map $d^i : [n-1] \to [n]$ in Δ induces a morphism of simplicial sets $d^i : \Delta[n-1] \to \Delta[n]$. Its image is a simplicial subset of $\Delta[n]$ called the ith face of $\Delta[n]$ and denoted by $\partial_i\Delta[n]$. The boundary $\partial\Delta[n] \subset \Delta[n]$ is the simplicial subset $\bigcup_{0\leq i\leq n} \partial_i\Delta[n]$.

The Yoneda lemma 2.35 yields bijections $\mathrm{Hom}_{s\mathrm{Set}}(\Delta[n], X) \simeq X_n$ for all simplicial sets X. The terminology 'standard n-simplex' for $\Delta[n]$ and 'face' for $\partial_i\Delta[n]$ is justified by the following geometrical interpretation.

Geometrical interpretation 3.3. Let $(v_0, \dots, v_n)$ be the canonical basis of $\mathbb{R}^{n+1}$. The geometric n-simplex Δ^n is the subspace of $\mathbb{R}^{n+1}$ which is the convex hull of the $n+1$ points v_i. The v_i are the vertices of Δ^n. For each morphism $f : [n] \to [m]$ in Δ we can define affine maps $f : \Delta^n \to \Delta^m$ by letting $f(v_k) = v_{f(k)}$. This yields an isomorphism between the full subcategory of sSet with objects the standard n-simplices onto the category whose objects are the geometric n-simplices, and whose morphisms are the affine maps sending vertices to vertices and preserving the order of the vertices.

Topological spaces provide examples of simplicial sets. Let T be a topological space. The *singular simplicial set of* T is the simplicial set $S(T)$ with $S(T)_n = \mathrm{Hom}_{\mathrm{Top}}(\Delta^n, T)$. The face operator $d_i : S(T)_n \to S(T)_{n-1}$ is induced by precomposing an n-simplex $\sigma : \Delta^n \to T$ by the affine map $d^i : \Delta^{n-1} \to \Delta^n$ (cf. Example 3.3). The degeneracy operator $s_i : S(T)_n \to S(T)_{n+1}$ is obtained by precomposing a simplex by the affine map $s^i : \Delta^{n+1} \to \Delta^n$. The following exercise shows that not all simplicial sets can be obtained as singular simplicial sets of topological spaces.

Exercise 3.4. Let $\Lambda^i[n]$ be the simplicial subset of $\Delta[n]$ obtained as the boundary of $\Delta[n]$ with the ith face removed:

$$\Lambda^i[n] = \textstyle\bigcup_{k\neq i} \partial_k\Delta[n] .$$

1. Show that the singular simplicial set of a topological space T satisfies the following 'Kan condition'. For all $n \geq 0$, for all $0 \leq i \leq n$ and for all morphisms $\sigma : \Lambda^i[n] \to X$, there exists morphisms $\overline{\sigma} : \Delta[n] \to X$ whose restriction to $\Lambda^i[n]$ equals σ.
2. Show that $\Delta[n]$ and $\Lambda^i[n]$ do not satisfy the Kan condition.

Categories provide another source of simplicial sets. Let $\mathcal{C}$ be a small category. An n-chain of composable morphisms is a chain of n morphisms

$$c_0 \xleftarrow{f_1} c_1 \xleftarrow{f_2} \cdots \xleftarrow{f_n} c_n \;.$$

We shall denote such a chain as a n-tuple $(f_1, \ldots, f_n)$. The *nerve* of $\mathcal{C}$ is the simplicial set $B\mathcal{C}$ defined as follows. $B\mathcal{C}_0$ is the set of objects of $\mathcal{C}$, $B\mathcal{C}_1$ is the set of morphisms of $\mathcal{C}$, and more generally for all $n \geq 1$, $B\mathcal{C}_n$ is the set of n-chains of composable morphisms. For $n = 1$, the face operators $d_0, d_1 : B\mathcal{C}_1 \to B\mathcal{C}_0$ are defined by

$$d_0 f = \mathrm{source}(f)\,, \qquad d_1(f) = \mathrm{target}(f)\;.$$

For $n \geq 2$, the face operators are given by dropping or composing morphisms:

$$d_i(f_1, \ldots, f_n) = \begin{cases} (f_2, \ldots, f_n) & \text{if } i = 0, \\ (f_1, \ldots, f_i f_{i+1}, \ldots, f_n) & \text{if } 0 < i < n, \\ (f_1, \ldots, f_{n-1}) & \text{if } i = n. \end{cases}$$

The degeneracy operators s_i are given by inserting the identity morphism in ith position. The following exercise characterizes the simplicial sets obtained as nerves of categories, and shows that the theory of (small) categories may be seen as a special case of the theory of simplicial sets.

Exercise 3.5. Let us denote by Cat the category whose objects are the small categories and whose morphisms are the functors.

1. Show that the nerve defines a functor $B : \mathrm{Cat} \to s\mathrm{Set}$.
2. The edge $[i, i+1]$ of $\Delta[n]$ is the image of $\Delta[1]$ by the morphism $\Delta[1] \hookrightarrow \Delta[n]$ induced by the map $[1] \to [n]$, $x \mapsto i + x$. The backbone of $\Delta[n]$ is the simplicial subset $[0,1] \cup \cdots \cup [n-1, n]$. Show that a simplicial set X is isomorphic to the nerve of a category if and only if it satisfies the following 'backbone condition'. For all morphisms of simplicial sets $\sigma : [0,1] \cup \cdots \cup [n-1, n] \to X$ there exists a unique simplex $\overline{\sigma} : \Delta[n] \to X$ whose restriction to the backbone of $\Delta[n]$ equals σ.
3. Show that B induces an equivalence of categories between Cat and the full subcategory of simplicial sets satisfying the 'backbone condition'.

3.1.4. Examples of simplicial R-modules and chain complexes. Let R be a ring. If X is a set, we denote by RX the free R-module with basis X. This yields a 'free R-module' functor

$$\begin{array}{rcl} \mathrm{Set} & \to & R\text{-Mod} \\ X & \mapsto & RX\,. \end{array}$$

Interesting examples of simplicial R-modules are provided by applying the free R-module functor to simplicial sets. Before giving illustrations of this, we need to introduce one more definition, relating simplicial R-modules to complexes of R-modules.

Definition 3.6. If M is a simplicial R-module, we define the associated chain complex $\mathcal{C}M$ by letting $(\mathcal{C}M)_n = M_n$ and the differential $d : (\mathcal{C}M)_n \to (\mathcal{C}M)_{n-1}$ is the sum $\sum_{i=0}^{n}(-1)^i d_i$. This defines a functor:

$$\mathcal{C} : s(R\text{-Mod}) \to \mathrm{Ch}_{\geq 0}(R\text{-Mod}) .$$

Exercise 3.7. Check that $\mathcal{C}M$ is indeed a chain complex.

Many of the most usual chain complexes are obtained from simplicial R-modules in this way. For example, let X be a topological space and let $S(X)$ be its singular simplicial set. We make it into a chain complex of R-modules $RS(X)$. The associated chain complex $\mathcal{C}RS(X)$ is the usual singular chain complex of X [Hat, Chap. 2], which computes the singular homology $H_*(X,R)$ of X with coefficients in R. Another example is provided by the following exercise.

Exercise 3.8. Let G be a discrete group, considered as a category with one object (cf. Section 2.5.1), and let BG be its nerve as defined in Exercise 3.5. Show that the chain complex of $\Bbbk$-modules $\mathcal{C}\Bbbk BG$ is equal to the bar complex $B(\Bbbk^{\mathrm{triv}}, \Bbbk G, \Bbbk^{\mathrm{triv}})$ defined in Section 2.3.3.

The chain complex $\mathcal{C}M$ admits the following variant, which is a smaller chain complex.

Definition 3.9. Let M be a simplicial R-module. The *normalized chain complex* $\mathcal{N}M$ is the subcomplex of $\mathcal{C}M$ defined by $(\mathcal{N}M)_n = \bigcap_{0\leq i<n} \ker d_i$, with differential $d = (-1)^n d_n : (\mathcal{N}M)_n \to (\mathcal{N}M)_{n-1}$. This defines a functor:

$$\mathcal{N} : s(R\text{-Mod}) \to \mathrm{Ch}_{\geq 0}(R\text{-Mod}) .$$

Proposition 3.10. *The normalized chain complex $\mathcal{N}M$ is a direct summand of the complex $\mathcal{C}M$. Moreover the inclusion $\mathcal{N}M \hookrightarrow \mathcal{C}M$ is a homotopy equivalence.*

Proof. Let us denote by DM the graded subobject of $\mathcal{C}M$ generated by the degeneracy operators, that is $(DM)_0 = 0$ and $(DM)_n = \sum_{0\leq i\leq n-1} s_i(M_{n-1})$. The simplicial identities imply that $(DM)_n \oplus (\mathcal{N}M)_n = (\mathcal{C}M)_n$, and that the differential of $\mathcal{C}M$ sends $(DM)_n$ into $(DM)_{n-1}$. Hence $\mathcal{C}M$ decomposes as the direct sum of complexes $\mathcal{C}M = \mathcal{N}M \oplus DM$.

To prove the homotopy equivalence $\mathcal{N}M \hookrightarrow \mathcal{C}M$, it suffices to prove that DM is homotopy equivalent to the zero complex, i.e., that Id_{DM} is homotopic to zero. For this, we write DM as an increasing union of subcomplexes

$$0 = D_{-1}M \subset D_0M \subset D_1M \subset \cdots \subset D_kM \subset \cdots \subset DM .$$

where $(D_kM)_n = \sum_{0\leq i\leq \min(k,n-1)} s_i(M_{n-1})$ for $n \geq 1$ (we use the simplicial identities to check that the D_kM are indeed subcomplexes of DM). We define $t_k : (DM)_n \to (DM)_{n+1}$ by $t_k(m) = (-1)^k s_k(m)$ if $k \leq \deg(m)$, 0 if $k > \deg(m)$, and we define

$$f_k := \mathrm{Id}_{DM} - dt_k - t_k d ,$$

where d is the differential of DM. By its construction, $f_k : DM \to DM$ is a chain map, which is homotopic to Id_{DM} (via the homotopy t_k), and which coincides with

Id_{DM} in low degrees: $f_k(m) = m$ if $\deg(m) < k$. The simplicial identities imply that for all $k \geq 0$:

$$f_k(D_kM) \subset D_{k-1}M\ , \quad f_k(D_jM) \subset D_jM \text{ if } j < k. \qquad (*)$$

Now we form the composite $f = f_0 f_1 f_2 \cdots f_k \cdots$. This composition is finite in each degree (since only a finite number of f_ks are not equal to the identity in a given degree), and is equal to the zero map by $(*)$. Now each f_k is homotopic to the identity, hence so is the composite f. □

3.1.5. Simplicial sets and algebraic topology. We now make a very quick digression on the relations between simplicial sets and algebraic topology. Although this section is not needed to understand the concept of derived functors of non-additive functors, it explains some notations and definitions of simplicial R-modules which originate from algebraic topology.

First of all, **simplicial sets are a combinatorial model for topological spaces.** To be more specific, from a simplicial set X we can construct a topological space $|X|$, the *realization of* X, by the following recipe. Consider each X_n as a discrete space, and recall the geometric n-simplex Δ^n described in the geometrical interpretation 3.3. Then $|X|$ is the topological space defined as the quotient

$$\Big(\bigsqcup_{n\geq 0} X_n \times \Delta^n\Big)/\sim\,,$$

where $\sim$ is the equivalence relation generated by $(x, d_i(\sigma)) \sim (d^i(x), \sigma)$ for $x \in \Delta^n$ and $\sigma \in X_{n+1}$ and $(x, s_i(\sigma)) \sim (s^i(x), \sigma)$, for $x \in \Delta^n$ and $\sigma \in X_{n-1}$ (and the affine maps $d^i : \Delta^n \to \Delta^{n+1}$ and $s^i : \Delta^n \to \Delta^{n-1}$ are the ones determined by $d^i : [n] \to [n+1]$ and $s^i : [n] \to [n-1]$ as explained in the geometrical interpretation 3.3). If Top denotes the category of topological paces and continuous maps, we have a realization functor:

$$|-| : s\mathrm{Set} \to \mathrm{Top}\,.$$

Among the important properties of the realization functor is the fact that the realization $|S(T)|$ of the singular simplicial set of a topological space T is always weakly homotopy equivalent to T itself (for more complete statements, we refer the reader, e.g., to [GJ, Chap. I, Thm. 11.4] or [Hov, Thm. 3.6.7]). This means that all the homotopy-theoretic information (like its homotopy groups or its singular homology groups) of the topological space T is encoded in the simplicial set $S(T)$.

In particular, **the classical invariants of topological spaces have their combinatorial counterpart in the world of simplicial sets.**

1. We can define the homology of a simplicial set X with coefficient in a ring R as the homology of the complex of R-modules $\mathcal{C}RX$. The homology of X is then isomorphic to the singular homology of its realization $|X|$ (when $X = S(T)$, this was already observed in Section 3.1.4).
2. The homotopy groups $\pi_n(X, x)$ of a simplicial set X with basepoint $x \in X_0$ can also be defined, at least when X is *fibrant*, that is when X satisfies the 'Kan condition' of Exercise 3.4 (the exercise shows that $S(T)$ is fibrant). We

refer the reader to [Wei, Def. 8.3.1], [GJ, Chap. I.7] of [Hov, Def. 3.4.4] for the definition. As for homology, one has an isomorphism $\pi_n(X,x) \simeq \pi_n(|X|,x)$.

Now a simplicial R-module M can be considered as a simplicial set by forgeting that the M_n are R-module, and just considering the underlying sets. This yields a a forgetful functor

$$s(R\text{-Mod}) \to s\text{Set} .$$

In particular, we can consider the homotopy groups of a simplicial R-module M, i.e., the homotopy groups of the underlying simplicial set. In this situation the combinatorial definition of homotopy groups specializes as $\pi_n(M,0) = H_n(\mathcal{N}M)$. This is why the homology groups of the (normalized) chain complex associated to a simplicial R-module M is commonly denoted by $\pi_n(M)$.

We can also define simplicial homotopies between morphisms of simplicial R-modules. This yields Definition 3.12 below. In particular, simplicially homotopic maps $f,g : M \to N$ yield homotopic continuous maps $|f|,|g| : |M| \to |N|$ after realization.

A topological space T is called an Eilenberg–Mac Lane space of type $K(A,n)$, with $n \geq 1$ and A an abelian group, if its homotopy groups are all zero except $\pi_n(T,*) \simeq A$. As another example of application of the viewpoint above, we call 'Eilenberg–Mac Lane space of type $K(A,n)$' a simplicial abelian group M whose homotopy groups are all zero, except $\pi_n(M) = A$. The Dold–Kan correspondence provides a way to build such Eilenberg–Mac Lane spaces, see Remark 3.16 below.

3.2. Derived functors of non-additive functors

Let $F : R\text{-Mod} \to S\text{-Mod}$ be a functor. If F is right exact we can derive it by using the classical theory of derived functors recalled in Section 2. The left derived functors of F are then defined by:

$$L_iF(M) = H_i(F(P^M)) , \quad L_iF(f) = H_i(F(\overline{f})) ,$$

where P^M is a projective resolution of the S-module M, and $\overline{f} : P^M \to P^N$ is a lifting of $f : M \to N$ at the level of the projective resolutions. The chain map $\overline{f}$ is only defined up to homotopy, but the definition is well founded because F sends homotopic chain maps to homotopic chain maps. This fact relies on the crucial fact that, being right exact, F is actually an additive functor, cf. Exercise 2.3.

Now if F is not additive, it does not send homotopic chain maps to homotopic chain maps (hence homotopy equivalences to homotopy equivalences), as the following example shows it. So we cannot use the formulas above to define its derived functors.

Example 3.11. Let $\Bbbk$ be a field, and let $\Lambda^2 : \Bbbk\text{-Vect} \to \Bbbk\text{-Vect}$ be the functor which sends a $\Bbbk$-vector space to its second exterior power. Let P and Q be the projective resolutions of the dimension one vector space respectively defined by: $P_i = 0$ if $i \neq 0$ and $P_0 = \Bbbk$; $Q_i = 0$ if $i > 1$, $Q_1 = \Bbbk$, $Q_0 = \Bbbk^2$ and $d : Q_1 \to Q_0$ sends λ to $(\lambda, 0)$. By Lemma 2.8, P and Q are homotopy equivalent. However $H_0(\Lambda^2(P)) = 0$ and $H_0(\Lambda^2(Q)) = \Bbbk$.

The suitable generalization of derived functors for arbitrary functors F : R-Mod $\to$ S-Mod was found by Dold and Puppe [DP], and relies on the Dold–Kan correspondence.

3.2.1. The Dold–Kan correspondence. We have seen in Definition 3.9 the normalized chain functor

$$\mathcal{N} : s(R\text{-Mod}) \to \mathrm{Ch}_{\geq 0}(R\text{-Mod}) .$$

We first establish a fundamental property of this functor, namely the preservation of homotopies.

Definition 3.12. Two morphisms of simplicial R-modules $f, g : M \to N$ are simplicially homotopic if there are morphisms $h_i : M_n \to N_{n+1}$ for $0 \leq i \leq n$ satisfying the following identities:

$$\begin{aligned} d_0 h_0 &= f, \\ d_{n+1} h_n &= g, \\ d_i h_j &= \begin{cases} h_{j-1} d_i & \text{if } i < j \\ d_i h_{i-1} & \text{if } i = j \neq 0 \\ h_j d_{i-1} & \text{if } i > j+1 \end{cases}, \\ s_i h_j &= \begin{cases} h_{j+1} s_i & \text{if } i \leq j \\ h_j s_{i-1} & \text{if } i > j \end{cases}. \end{aligned}$$

Proposition 3.13. *Let $f, g : M \to N$ be two simplicially homotopic morphisms. Then the induced chain maps at the level of normalized chain complexes $\mathcal{N}(f)$, $\mathcal{N}(g) : \mathcal{N}M \to \mathcal{N}N$ are chain homotopic.*

Proof. Let $h_i : M_n \to N_{n+1}$, $0 \leq i \leq n$ be the maps defining the homotopy from f to g (so $d_0 h_0 = f$, $d_{n+1} h_n = g$). We let $t_n : M_n \to N_{n+1}$ be the map defined by

$$t_n(m) = \sum_{j=0}^{n} (-1)^j (h_j(m) - f s_j(m)) .$$

We denote by $t_n^{\leq i}(m)$ and $t_n^{\geq i}$ the partial sums obtained when the summation index j varies from 0 to i, resp. from i to n (so $t_n = t_n^{\leq i} + t_n^{\geq i+1}$). The definition of simplicial homotopies imply that:

$$\begin{aligned} d_0 t_n &= -t_{n-1} d_0 \\ d_i t_n &= t_{n-1}^{\leq i-2} d_{i-1} - t_{n-1}^{\geq i} d_i \\ d_{n+1} t_n &= t_{n-1} d_n + (-1)^n (g - f) . \end{aligned}$$

As a consequence, t_n maps $(\mathcal{N}M)_n$ to $(\mathcal{N}N)_{n+1}$, and the differential d of $\mathcal{N}M$ satisfies $t_n d + d t_{n+1} = f - g$. □

Remark 3.14. Let us denote by $\pi_n(M)$ the nth homotopy group of a simplicial R-module M, that is the nth homology group of its normalized chain complex $\mathcal{N}M$. Proposition 3.13 shows that simplicially homotopic maps induce the same map at the level of homotopy groups. This should not surprise the reader after the discussion of Section 3.1.5

Theorem 3.15 (The Dold–Kan correspondence). *The normalized chain functor $\mathcal{N}$ is an equivalence of categories. Moreover, it has an inverse $\mathcal{K} : \mathrm{Ch}_{\geq 0}(R\text{-Mod}) \to s(R\text{-Mod})$ which preserves homotopies.*

The proof of this theorem is quite long and we will skip it. The reader can consult [DP, Section 3], [GJ, Chap. III.2] or [Wei, Section 8.4] for a proof. We just insist on the fact that the functor $\mathcal{K}$ has a very explicit combinatorial expression. For example, let $M[1]$ denote the chain complex which is zero except in degree 1, where it equals the R-module M. Then the value of $\mathcal{K}$ on the complex $M[1]$ is just the simplicial R-module with $\mathcal{K}(M[1])_n = M^{\oplus n}$, the face operators $d_i : M^{\oplus n} \to M^{\oplus n-1}$ are defined by

$$\begin{aligned} d_0(x_1, \ldots, x_k) &= (x_2, \ldots, x_n) \ , \\ d_i(x_1, \ldots, x_k) &= (x_1, \ldots, x_i + x_{i+1}, \ldots, x_n) \text{ if } 0 < i < k \ , \\ d_n(x_1, \ldots, x_k) &= (x_1, \ldots, x_{n-1}) \ , \end{aligned}$$

and the degeneracy operators s_i insert a zero in ith position.

Remark 3.16. Let $M[n]$ denote the chain complex which is zero except in degree n, where it equals the R-module M. Then $\mathcal{K}(M[n])$ is a simplicial Eilenberg–Mac Lane space, i.e., a simplicial R-module whose homotopy groups are zero, except $\pi_n(\mathcal{K}(M[n])) = M$. In particular, the functor $\mathcal{K}$ may be used as a constructor of Eilenberg–Mac Lane spaces.

3.2.2. Derived functors. Let $F : R\text{-Mod} \to S\text{-Mod}$ be a functor. As already observed, the induced functor $F : \mathrm{Ch}_{\geq 0}(R\text{-Mod}) \to \mathrm{Ch}_{\geq 0}(S\text{-Mod})$ by applying F degreewise does not preserve chain homotopies if F is not additive. The key observation is that even if F is not additive, the induced functor $F : s(R\text{-Mod}) \to s(S\text{-Mod})$ does preserve simplicial homotopies. Indeed, simplicial homotopies do not involve additions in their definition. Since the Dold–Kan correspondence preserves homotopies, the following composite functor preserves chain homotopies:

$$\mathrm{Ch}_{\geq 0}(R\text{-Mod}) \xrightarrow[\sim]{\mathcal{K}} s(R\text{-Mod}) \xrightarrow{F} s(S\text{-Mod}) \xrightarrow[\sim]{\mathcal{N}} \mathrm{Ch}_{\geq 0}(S\text{-Mod}) \ .$$

So, we can make the following well-founded definition.

Definition 3.17 ([DP]). Let $F : R\text{-Mod} \to S\text{-Mod}$ be a functor, and let n be a nonnegative integer. The ith derived functor of F with height n is defined by

$$L_iF(M; n) = H_i(\mathcal{N}F\mathcal{K}(P^M[n])) \ , \quad L_iF(f; n) = H_i(\mathcal{N}F\mathcal{K}(\overline{f})[n]) \ ,$$

where P^M is a projective resolution of the S-module M, $\overline{f} : P^M \to P^N$ is a lifting of $f : M \to N$ at the level of the projective resolutions, and $[n]$ refers to the n-fold suspension functor $[n] : \mathrm{Ch}_{\geq 0}(R\text{-Mod}) \to \mathrm{Ch}_{\geq 0}(R\text{-Mod})$.

The following proposition shows that Definition 3.17 is a generalization of the classical definition of derived functors of semi-exact functors.

Proposition 3.18. *Let $F : R\text{-Mod} \to S\text{-Mod}$ be a right-exact functor and let $n \geq 0$. For all $i \geq 0$, there is a natural isomorphism (with the convention that derived functors with negative indexes are zero):*

$$L_iF(M;n) \simeq L_{i-n}F(M)\,.$$

Proof. Since F is right exact, it is additive. In particular, if X is a simplicial R-module, the complexes $F(\mathcal{C}X)$ and $\mathcal{C}F(X)$ are equal (by definition, $(F(\mathcal{C}X))_n = F(X_n) = (\mathcal{C}F(X))_n$ and by additivity, $F(\sum(-1)^i d_i) = \sum(-1)^i F(d_i)$, hence $F(\mathcal{C}X)$ and $\mathcal{C}F(X)$ have the same differentials). Additivity of F also ensures that F preserves homotopy equivalences and commutes with suspension. Thus we have a chain of homotopy equivalences of chain complexes:

$$\begin{aligned}\mathcal{N}F\mathcal{K}(P^M[n]) \simeq \mathcal{C}F\mathcal{K}(P^M[n]) = F\mathcal{C}\mathcal{K}(P^M[n])\,,\\ \simeq F\mathcal{N}\mathcal{K}(P^M[n]) \simeq F(P^M[n]) = F(P^M)[n]\,.\end{aligned}$$

Taking the homology of the corresponding complexes, we get the result. □

The definition of derived functors can be generalized so that the derived functors are evaluated on complexes of R-modules, instead of complexes of the form $M[n]$. To be more specific, if C is a complex of R-modules, we denote by $LF(C)$ the simplicial S-module

$$LF(C) := F(\mathcal{K}(C'))\,,$$

where C' is a complex of projective R-modules which is quasi-isomorphic to C. Such a C' always exists, and is unique up to a chain homotopy equivalence [Wei, Section 5.7]. Recall that the homotopy groups of a simplicial R-module M are defined by $\pi_i M := H_i(\mathcal{N}M)$. Then the values of the left derived functor of F on C are just the homotopy groups

$$L_iF(C) := \pi_i LF(C)\,.$$

In particular, the left derived functors with height n of Definition 3.17 are given by $L_iF(M,n) := \pi_i LF(M[n])$.

Remark 3.19. Quillen has put the definition of Dold and Puppe into perspective, in the framework of his homotopical algebra [Q]. There is a model structure on $s(R\text{-mod})$ where the weak equivalences are the morphisms $f : M \to N$ inducing isomorphisms on the level of homotopy groups, and cofibrant objects are simplicial R-modules M with M_n R-projective for all $n \geq 0$. We denote by $\mathbf{D}_{\geq 0}(R\text{-mod})$ the localization of $s(R\text{-Mod})$ at weak equivalences. By the Dold–Kan correspondence, it is isomorphic to the localization of the category $\mathrm{Ch}_{\geq 0}(R\text{-mod})$ at quasi-isomorphisms. So we can indifferently consider the objects of $\mathbf{D}_{\geq 0}(R\text{-mod})$ as chain complexes or as simplicial R-modules. If $F : R\text{-Mod} \to S\text{-Mod}$ is a functor, the induced functor $F : s(R\text{-Mod}) \to s(S\text{-Mod})$ admits a Quillen derived functor

$$LF : \mathbf{D}_{\geq 0}(R\text{-mod}) \to \mathbf{D}_{\geq 0}(S\text{-mod})$$

whose value on a complex C is precisely the simplicial object $LF(C)$ defined above.

3.2.3. Examples of derived functors. For the sake of concreteness, we finish the section by giving two topological situations where derived functors of non-additive functors appear.

First, derived functors of non-additive functors are related to the singular homology of Eilenberg–Mac Lane spaces. To be more specific, let $\mathbb{Z}A$ denote the group ring of an abelian group A. We can consider the group ring as a functor

$$\mathbb{Z}- : \mathrm{Ab} \to \mathrm{Ab}\,.$$

Then it follows from the discussion of Section 3.1.5 and Remark 3.16 that the derived functors of the group ring functor with height n compute the singular homology of the Eilenberg–Mac Lane space $K(A,n)$; there is an isomorphism of functors of the abelian group A:

$$L_i\mathbb{Z}(A;n) \simeq H_i(K(A,n);\mathbb{Z})\,.$$

The case $n=1$ might be particularly interesting, since the singular homology of $K(A,1)$ is isomorphic to the homology of an abelian group A [Br, Chap. II, Section 4]. In particular, there is an isomorphism of functors of the abelian group A:

$$L_i\mathbb{Z}(A;1) \simeq H_i(A;\mathbb{Z}) = \mathrm{Tor}_i^{\mathbb{Z}A}(\mathbb{Z}^{\mathrm{triv}},\mathbb{Z}^{\mathrm{triv}})\,.$$

Let us denote by $S^n(A)$ the nth symmetric power of an abelian group A. We can consider the nth symmetric power as a functor

$$S^n : \mathrm{Ab} \to \mathrm{Ab}\,.$$

The symmetric power has an analogue for topological spaces; the nth symmetric product $SP^n(X)$ of a topological space X is the space of orbits of $X^{\times n}$ under the action of the symmetric group $\mathfrak{S}_n$ acting on $X^{\times n}$ by permuting the factors. Dold proved [Do] that if X is a CW-complex, the singular homology of $SP^n(X)$ can be computed in terms of the homology of X and the derived functors of S^n (evaluated on the graded object $H_*(X;\mathbb{Z})$, considered as a complex with trivial differential):

$$\pi_i LS^n(H_*(X;\mathbb{Z})) \simeq H_i(SP^n(X);\mathbb{Z})\,.$$

4. Spectral sequences

Spectral sequences are a powerful tool to compute derived functors. They can be thought of as an optimal way to organize long exact sequences in computations.

4.1. A quick overview

Let $\mathcal{A}$ be an abelian category, e.g., the category of R-modules. Spectral sequences appear in the following typical situation. We want to compute a graded object in $\mathcal{A}$, which we denote by K_* (e.g., the singular homology of a topological space, some derived functors, etc.). We don't know K_* but we can break it into smaller pieces which we understand. Roughly speaking, a spectral sequence is a kind of algorithm, which takes the pieces of K_* as an input, and which computes K_*. Let us state some formal definitions.

Definition 4.1. Let r_0 be a positive integer.

A homological spectral sequence in $\mathcal{A}$ is a sequence of chain complexes in $\mathcal{A}$, $(E_*^r, d^r)_{r \geq r_0}$, such that for all $r \geq r_0$, $E_*^{r+1} = H(E_*^r)$.

A cohomological spectral sequence in $\mathcal{A}$ is a sequence of cochain complexes in $\mathcal{A}$, $(E_r^*, d_r)_{r \geq r_0}$ such that for all $r \geq r_0$, $E_{r+1}^* = H(E_r^*)$.

Cohomology spectral sequences can be converted into homology spectral sequences (and vice versa) by the usual trick on complexes $E_r^i = E_{-i}^r$. So we concentrate on homological spectral sequences and leave to the reader the translation for cohomological spectral sequences. The term E_*^r of a homological spectral sequence is called the rth page of the spectral sequence, and d^r is called the differential of the rth page. The term $E_*^{r_0}$ is called the initial term of the spectral sequence.

In the sequel, we shall often omit to mention the category $\mathcal{A}$ in which the spectral sequence lives. Thus spectral sequences in $\mathcal{A}$ will simply called spectral sequences, etc. This will cause no confusion since the category $\mathcal{A}$ is transparent in all the definitions, we only need the fact that it is abelian.

Definition 4.2. A spectral sequence $(E_*^r, d^r)_{r \geq 0}$ is stationary if for all $k \in \mathbb{Z}$, there is an integer $r(k)$ such that for $r \geq r(k)$ we have $E_k^r = E_k^{r(k)}$. In that case we define its E^∞-page by the formula $E_k^\infty := E_k^{r(k)}$ for all $k \in \mathbb{Z}$.

The following definition of convergence of a spectral sequence is not the most general one[6], but it is sufficient for our applications.

Definition 4.3 (Convergence). Let K_* be a graded object. We say that a spectral sequence $(E_*^r, d^r)_{r \geq 0}$ converges to K_* if (i) the spectral sequence is stationary, and (ii) there exists a filtration of each K_k: $\cdots \subset F_{p-1}K_k \subset F_pK_k \subset \cdots \subset K_k$, which is exhaustive: $\bigcup_p F_pK_k = K_k$, Hausdorff: $\bigcap_p F_pK_k = 0$, and there is a graded isomorphism $\mathrm{Gr}(K_*) \simeq E_*^\infty$.

Typical theorems for spectral sequences assert the existence of a spectral sequence with an explicit initial term, say E_*^2, converging to what we want to compute: K_*. When such a spectral sequence exists, if we know the initial page, we can run the algorithm provided by the spectral sequence (i.e., compute the successive homologies) to obtain the E^∞-page, hence $\mathrm{Gr}(K_*)$, hence K_*.

There are many other possible uses of spectral sequences. For example, we can use spectral sequences backwards: we know the abutment K_* and we want to compute the initial object E_*^2, so we use the spectral sequence to obtain information on E_*^2. Spectral sequences can also be used to propagate properties. For example, let us take the case of a spectral sequence of abelian groups (E_*^r, d^r). Assume that the graded abelian group E_*^2 is finitely generated in each degree. Then all its subquotients are finitely generated in each degree, so E_*^∞ is finitely generated in each degree. Thus we deduce that the abutment K_* is also finitely generated in each degree.

[6]It is possible to define convergence (in particular the E^∞-term) for non-stationnary spectral sequences. We refer the reader to [Boa] for the most general statements.

4.2. Bigraded spectral sequences

In practice, most spectral sequences are bigraded. In this section, we rewrite the definitions of the previous section in this more complicated setting.

Definition 4.4. A homological spectral sequence is a sequence $(E^r_{*,*}, d^r)_{r\geq r_0}$ of bigraded objects $E^r_{*,*}$, equipped with differentials d^r of bidegree $(-r, r-1)$[7], i.e., d_r restricts to maps $d_r : E^r_{p,q} \to E^r_{p-r,q+r-1}$. The terms of the spectral sequence are required to satisfy $E^{r+1}_{*,*} = H(E^r_{*,*})$.

Let $E^r_k = \bigoplus_{p+q=k} E^r_{p,q}$ be the summand of total degree k of the rth page. Then the differential of the spectral sequence lowers the total degree by 1. So bigraded spectral sequences are a refinement of Definition 4.1.

Bigraded spectral sequences are often depicted by a diagram of the following type (here we represent the second page, the summands $E^2_{p,q}$ of the second page are represented by dots, and the dots on the same dashed lines have the same total degree):

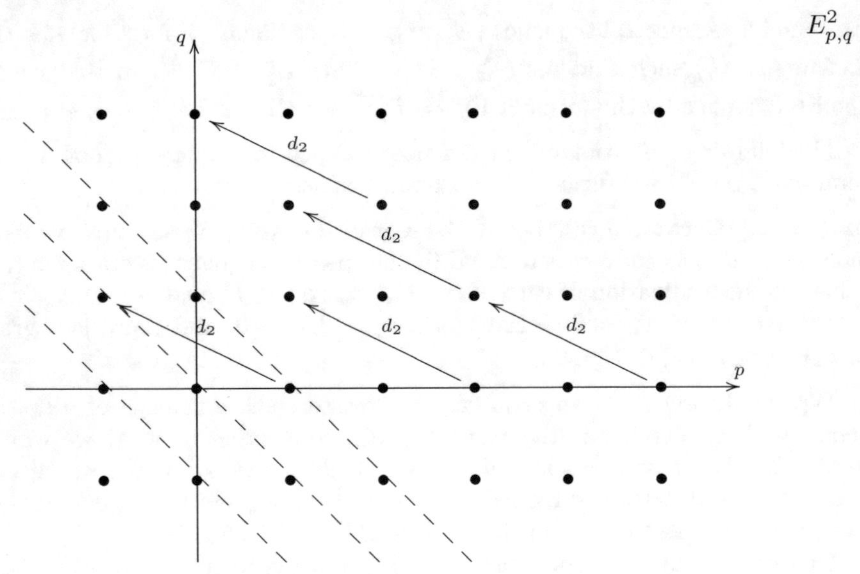

Definition 4.5. A spectral sequence is stationary if for all $p, q \in \mathbb{Z}^2$ there is an integer $r(p,q)$ such that $E^r_{p,q} = E^{r(p,q)}_{p,q}$ for $r \geq r(p,q)$. For a stationary spectral sequence, we define $E^\infty_{p,q} := E^r_{p,q}$ for $r \gg 0$.

Bigraded spectral sequences converge to *graded* objects. Roughly speaking, a convergent bigraded spectral sequence converges to K_* if for all k, K_k is filtered, and there is an isomorphism $E^\infty_k = \bigoplus_{p+q=k} E^\infty_{p,q} \simeq \mathrm{Gr}(K_k)$. Thus, only the total degree in the spectral sequence is really meaningful to recover the abutment. The

[7] Other conventions on the bidegrees of the differentials are possible, but this convention is the most common one.

bidegree should be considered as a technical refinement of the total degree, useful for intermediate computations[8]. Having this explained, we can now state the full definition of the convergence of a spectral sequence.

Definition 4.6. Let K_* be a graded object. A spectral sequence $(E^r_{*,*}, d^r)$ converges to K_* if (i) it is stationary, and (*ii*) for all k there is an exhaustive Hausdorff filtration:

$$0 = \left(\bigcap_p F_p K_k\right) \subset \cdots \subset F_{p-1}K_k \subset F_p K_k \subset \cdots \subset \left(\bigcup_p F_p K_k\right) = K_k \,,$$

together with isomorphisms: $E^\infty_{p,q} \simeq F_p K_{p+q} / F_{p-1} K_{p+q}$.

The convergence of a spectral sequence with initial page r_0 is often written in the symbolic way:

$$E^{r_0}_{p,q} \Longrightarrow K_{p+q} \,.$$

4.3. A practical example

The archetypal theorem for spectral sequences is the following theorem[9].

Theorem 4.7 (Leray–Serre). *Let $f : X \to B$ be a fibration between topological spaces. Assume that the base space B is simply connected, and denote by $F = f^{-1}(\{b\})$ the fiber over an element $b \in B$. There is a homological spectral sequence of abelian groups*

$$E^2_{p,q} = H_p(B, H_q(F)) \Longrightarrow H_{p+q}(X) \,.$$

In the statement, the spectral sequence is bigraded, and starts at the second page. The bidegree of the differential is not indicated (so the reader should think that the rth differential d^r has bidegree $(-r, r-1)$ as it is most often the case). The summands $E^2_{p,q}$ are nonzero only if $p \geq 0$ and $q \geq 0$: such a spectral sequence is called a *first quadrant* spectral sequence.

Thus, assume for example that we know $H_*(B)$ and $H_*(F)$, hence the bigraded abelian group $H_*(B, H_*(F))$. Then we can use the Leray–Serre spectral sequence as an algorithm to compute $H_*(X)$. Unfortunately, we immediately encounter three practical problems.

4.3.1. Problem 1: does the spectral sequence stop? To compute the E^∞-page of the Leray–Serre spectral sequence, we must start from the initial page, compute its homology $E^3_{p,q}$, then compute the homology $E^4_{p,q}$ of the third page, and so on. Since the spectral sequence is convergent, we know that for all indexes (p, q) the process stops at some point: $E^r_{p,q} = E^\infty_{p,q}$ for r big enough. But it may happen that the process does not stop for all the indices at the same time. This is a problem, because in

[8] In Definition 4.6, the bigrading on $E^\infty_{*,*}$ gives additional information on the isomorphism $E^\infty_* \simeq \mathrm{Gr}(K_*)$. But in many cases, this additional information is useless, so the reader can keep Definition 4.3 in mind.

[9] In this theorem, the letter H refers to the singular homology of a topological space. The reader unfamiliar with algebraic topology can think of a fibration $X \to B$ with fiber F as a 'topological space X which is an extension of B by F'. The hypothesis that B is simply connected is not necessary, but in the general case, the statement is slightly more complicated.

that case we need infinitely many successive computations to compute completely $E^{\infty}_{*,*}$ (doing these computations would exceed a mathematician's lifetime).

Definition 4.8. A spectral sequence $(E^r_{*,*}, d^r)$ stops at the sth page if for all $r \geq s$, the differential d^r is zero. (Thus $E^{\infty}_{*,*} = E^s_{*,*}$.)

In some cases, stopping follows from the shape of the initial page of the spectral sequence. For example, assume that F and B are topological spaces with bounded singular homology. Then $E^2_{p,q}$ might be non-zero only in some rectangle $0 \leq p \leq p_0$, $0 \leq q \leq q_0$. Thus, for all r big enough ($r \geq \min(p_0+1, q_0+2)$), either the target or the source of d^r is zero: the differentials d^r are 'too long'. So the spectral sequence stops at the $\min(p_0+1, q_0+2)$th page. Here is another situation where stopping follows from the shape of the spectral sequence.

Exercise 4.9. Let $(E^r_{p,q}, d^r)_{r\geq 2}$ be a homological spectral sequence. Assume that $E^r_{p,q} = 0$ if p is odd or if q is odd. Show that $E^2_{p,q} = E^{\infty}_{p,q}$.

Stopping can also occur under more subtle hypotheses.

Exercise 4.10. Let $(E^r_{p,q}, d^r)_{r\geq 2}$ be a spectral sequence. Assume that there is a graded ring R such that each $(E^r_{p,q}, d_r)$ is a complex of graded R-modules, and $E^2_{p,q}$ is a noetherian R-module. Prove that the spectral sequence stops. (Hint: consider the sequences of R-submodules of the second page:

$$B^2 \subset B^3 \subset \cdots \subset B^i \subset \cdots \quad \subset \quad \cdots \subset Z^i \subset \cdots \subset Z^3 \subset Z^2 ,$$

where $Z^2 = \ker d^2$, $B^2 = \operatorname{Im} d^2$ and the Z^i, B^i for $i > 2$ are defined inductively by $Z^{i+1} = \pi^{-1}(\ker d^{i+1})$, $B^{i+1} = \pi^{-1}(\operatorname{Im} d^{i+1})$ where π is the map $Z^i \twoheadrightarrow Z^i/B^i = E^{i+1}$.)

4.3.2. Problem 2: differentials are not explicit. Theorem 4.7 asserts that the spectral sequence exist, but does not give any formula for the differential! So, even if we know what the second page is, it is not possible to compute the third page in general. **It is the main and recurrent problem for spectral sequences**.

To bypass this problem, one can use additional algebraic structure on the spectral sequence (see Section 4.4). Alternatively, some differentials, or some algebraic maps related to the spectral sequence might have a geometric interpretation, which can be used to compute them. For example, the full version of the Leray–Serre spectral sequence comprises the following statement in addition to Theorem 4.7.

Theorem 4.11 (Leray–Serre – continued). *If $e_B : E^{\infty}_{p,0} \hookrightarrow E^2_{p,0}$ denotes the canonical inclusion, the composite*

$$H_p(X) \twoheadrightarrow F_pH_p(X)/F_{p-1}H_p(X) \simeq E^{\infty}_{p,0} \xrightarrow{e_B} E^2_{p,0} \simeq H_p(B, H_0(F)) \simeq H_p(B)$$

is equal to the map $H_p(f) : H_p(X) \to H_p(B)$. Similarly, if $e_F : E^2_{0,q} \twoheadrightarrow E^{\infty}_{0,q}$ denotes the canonical surjection, the composite

$$H_q(F) \simeq H_0(B, H_q(F)) \simeq E^2_{0,q} \xrightarrow{e_F} E^{\infty}_{0,q} \simeq F_0H_n(X) \hookrightarrow H_n(X)$$

equals the map $H_n(F) \to H_n(E)$ induced by the inclusion $F \subset X$.

The composites appearing in the second part of Theorem 4.11 are called the *edge maps*. Edge maps are a common feature of first quadrant spectral sequences, and the 'geometric' interpretation of edge maps can sometimes be used to compute some differentials. For example, let us assume that f has a section, i.e., there exists a map $s : B \to X$ such that $f \circ s = \mathrm{Id}_B$. Then $H_*(f)$ (hence e_B) is onto. Thus, e_B is an isomorphism. Therefore, all the elements of $E^2_{*,0}$ survive to the E^∞ page. That is, all the differentials starting from $E^2_{*,0}$ must be zero.

4.3.3. Problem 3: extension problems. Assume now that we have (finally) succeeded in computing $E^\infty_{*,*}$. We have not finished yet! Indeed the E^∞-page of the spectral sequence is isomorphic to $\mathrm{Gr}(H_*(X))$, not to $H_*(X)$. This can make a big difference. For example, if E^∞_k equals $\mathbb{Z}/2\mathbb{Z} \oplus \mathbb{Z}/2\mathbb{Z}$, then $H_k(X)$ could equal $\mathbb{Z}/2\mathbb{Z} \oplus \mathbb{Z}/2\mathbb{Z}$ or $\mathbb{Z}/4\mathbb{Z}$.

In the case of spectral sequences of $\mathbb{k}$-vector spaces (or more generally if the E^∞-page is a bigraded projective $\mathbb{k}$-module) this problem vanishes thanks to the following lemma.

Lemma 4.12. *Let M be a $\mathbb{k}$-module, equipped with a filtration*

$$0 = F_m M \subset \cdots \subset F_{p-1}M \subset F_p M \subset \cdots \subset \left(\bigcup\nolimits_{p\geq m} F_p M\right) = M\,.$$

Assume that $\mathrm{Gr}(M)$ *is a projective $\mathbb{k}$-module. Then there is an isomorphism of $\mathbb{k}$-modules $M \simeq \mathrm{Gr}(M)$.*

Proof. Let $0 \to M' \to M \to M'' \to 0$ be a short exact sequence of $\mathbb{k}$-modules. If M'' is projective, we can find a section of the map $M \to M''$ so M splits as a direct sum: $M \simeq M' \oplus M''$. We build the isomorphism $M \simeq \mathrm{Gr}(M)$ by iterative uses of this fact. □

To solve extension problems in the general case, we often use additional structure on the spectral sequence, or additional information on the abutment $H_*(X)$, obtained independently from the spectral sequence.

4.4. Additional structure on spectral sequences

Many spectral sequences bear more structure than what is stated in Definition 4.4. This additional structure is usually of great help in effective computations. In this section we present two properties among the most frequent and useful ones, namely algebra structures (the pages of the spectral sequences are bigraded algebras), and naturality (the spectral sequence depends functorially on the object it is built from).

4.4.1. Spectral sequences of algebras

Definition 4.13. Let $(E^r_{*,*}, d^r)$ be a spectral sequence of R-modules, converging to K_*. We say that it is a spectral sequences of algebras if the following conditions are satisfied.

(1) For all r there is a bigraded product $E^r_{p_1,q_1} \otimes_R E^r_{p_2,q_2} \to E^r_{p_1+p_2,q_2+q_2}$, satisfying a Leibniz relation

$$d^r(x_1 \cdot x_2) = d^r(x_1) \cdot x_2 + (-1)^{p_1+q_1} x_1 \cdot d^r(x_2) \quad \text{for } x_i \in E^r_{p_i,q_i} .$$

In particular $H(E^r_{*,*})$ is a bigraded algebra.

(2) The isomorphism $H(E^r_{*,*}) \simeq E^{r+1}_{*,*}$ is compatible with products.

(3) The abutment K_* is a filtered graded algebra (i.e., the filtration satisfies $F_p K_i \cdots F_q K_j \subset F_{p+q} K_{i+j}$) and the isomorphism $E^\infty_* \simeq \mathrm{Gr}(K_*)$ is an isomorphism of graded algebras.

Spectral sequences of algebras are interesting for many purposes. First, one might very well be interested in computing the graded R-algebra K_* rather than the graded R-module K_*. In that case, we need a spectral sequence of algebras to do the job. But even if we are not interested in the algebra structure of the abutment, the algebra structure on a spectral sequence $(E^r_{*,*}, d^r)$ is an extremely useful tool to compute the differentials. Indeed, by the Leibniz rule, it suffices to determine $d^r(x)$ for generators x of $E^r_{*,*}$ to completely determine d^r on $E^r_{*,*}$! Examples of spectral sequences of algebras are given in Section 4.5.

4.4.2. Naturality of spectral sequences. Let $\mathcal{A}$ be an abelian category. To define the category of homological spectral sequences in $\mathcal{A}$, we need to define morphisms of spectral sequences.

Definition 4.14. Let $(E^r_{*,*}, d^r)$ and $(E'^r_{*,*}, d'^r)$ be two spectral sequences converging to K_* and K'_* respectively. A morphism of spectral sequences is a sequence of bigraded maps $f^r : E^r_{*,*} \to E'^r_{*,*}$, which commute with the differentials: $d'^r(f(x)) = f(d^r(x))$, and such that $H(f^r) = f^{r+1}$. (In particular, the morphism induces a map $f^\infty : E^\infty_{*,*} \to E'^\infty_{*,*}$.)

From a practical point of view, morphisms of spectral sequences $f^r : E^r_{*,*} \to E'^r_{*,*}$ are very useful for explicit computations. Indeed, if the differentials in the spectral sequence $E^r_{*,*}$ are known, then one can use f^r to prove that $d'^r(x) = 0$, for x in the image of f^r, or to prove that some some x in the image of f^r are boundaries.

In practice, spectral sequences usually come from a functor

$$E : \mathcal{C} \to \text{Spectral sequences} ,$$

where $\mathcal{C}$ is a given category. One also says that the spectral sequence is natural with respect to the objects $C \in \mathcal{C}$. In this situation, morphisms in $\mathcal{C}$ induce maps of spectral sequences. The Leray–Serre spectral sequence is a typical example ($\mathcal{C}$ is the category of fibrations).

Theorem 4.15 (Leray–Serre – continued). *The Leray–Serre spectral sequence is natural with respect to the fibration f, and the convergence is natural with respect to f.*

The precise meaning of Theorem 4.15 is the following. If (g_B, g_X, g_F) is a morphism of fibrations, i.e., there is a commutative diagram:

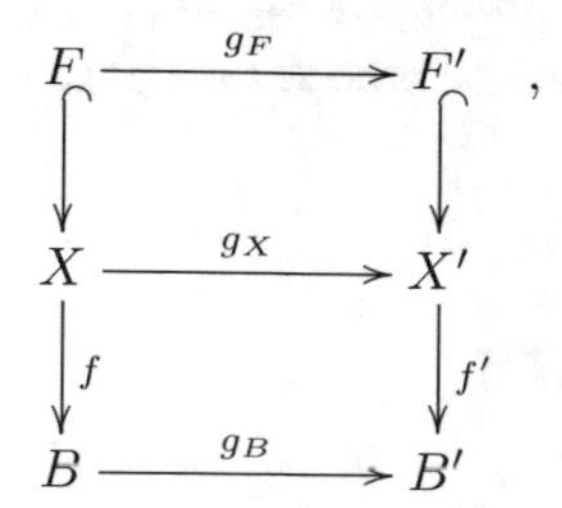

then (g_B, g_X, g_F) induces a morphism $g^r : E^r_{*,*} \to E'^r_{*,*}$ of spectral sequences between the two Leray–Serre spectral sequences. On the level of the second page, g^2 coincides with the morphism $H_*(g_B, H_*(g_F))$. The map $H_*(g_X)$ preserves the filtrations on the abutments and $\mathrm{Gr}(H_*(g_X))$ equals the map $g^\infty : E^\infty_{*,*} \to E'^\infty_{*,*}$.

4.5. Examples of spectral sequences

In this section we give some of the most common examples of spectral sequences. We shall not explain in details how these spectral sequences are constructed, since their construction is quite involved and usually useless for practical computations. We refer the reader to [Wei, Chap. 5], [ML, XI], [Ben, Chap. 3] or [MC] for more information on their construction.

4.5.1. Filtered complexes. Let C be a chain complex in an abelian category $\mathcal{A}$. A filtration of C is a family of subcomplexes $\cdots \subset F_pC \subset F_{p+1}C \subset \cdots$ of C. The filtration is *bounded below* if for all k, there is an integer p_k such that $F_{p_k}C_k = 0$. It is *exhaustive* if $\bigcup_{p\in\mathbb{Z}} F_pC = C$.

If C is a filtered complex, we denote by Gr_pC the quotient complex: $\mathrm{Gr}_pC = F_pC/F_{p-1}C$. The filtration of C induces a filtration on its homology, defined by

$$F_pH_*(C) := \mathrm{Im}\big(H_*(F_pC) \to H_*(C)\big) .$$

The following theorem provides a spectral sequence which recovers the homology of C from the homology of the complexes Gr_pC.

Theorem 4.16. *Let C be a filtered chain complex, whose filtration is bounded below and exhaustive. There is a homological spectral sequence*

$$E^1_{p,q}(C) = H_{p+q}(\mathrm{Gr}_pC) \Longrightarrow H_{p+q}(C) .$$

More explicitly, there are isomorphisms

$$E^\infty_{p,q}(C) \simeq F_pH_{p+q}(C)/F_{p-1}H_{p+q}(C) .$$

If $f : C \to C'$ is a chain map preserving the filtrations (i.e., it sends F_pC into F_pC'), then f induces a morphism of spectral sequences with $E^1_{p,q}(f) = H_{p+q}(\mathrm{Gr}_pf)$, and $E^\infty_{p,q}(f)$ coincides on the abutment with $\mathrm{Gr}_pH_{p+q}(f)$.

Example 4.17. Let us test Theorem 4.16 on the simplest example. Let $0 \to C' \to C \to C'' \to 0$ be a short exact sequence of chain complexes. This short exact sequence is equivalent to saying that C is filtered with $F_0C = C'$ and $F_1C = C$. Thus we have a spectral sequence whose first page is of the form:

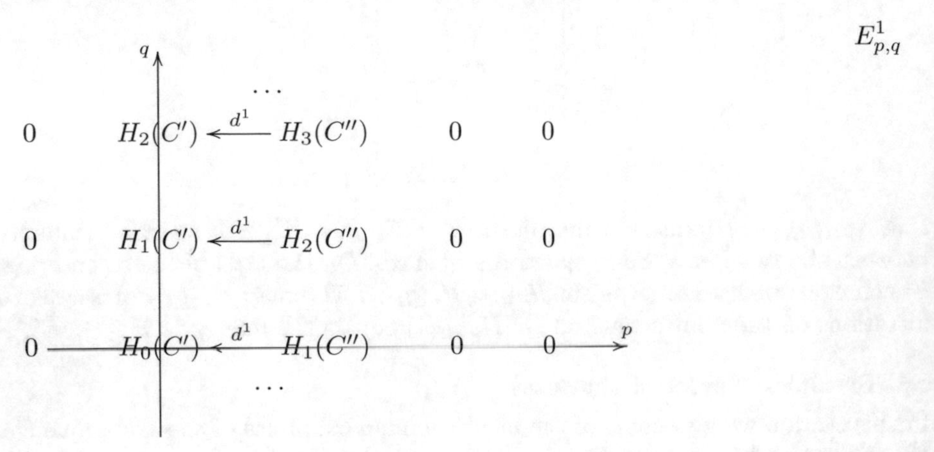

The shape of the E^1-page implies that the differentials d_r are zero for $r > 1$ (these differentials are 'too long': their source or their target must be zero), so $E^2_{p,q} = E^\infty_{p,q}$. As a result we have exact sequences:

$$0 \to E^\infty_{1,q} \to H_{q+1}(C'') \xrightarrow{d_1} H_q(C') \to E^\infty_{0,q} \to 0\,.$$

Now the convergence of the spectral sequence is equivalent to the data of short exact sequences:

$$0 \to E^\infty_{0,q+1} \to H_{q+1}(C) \to E^\infty_{1,q} \to 0\,.$$

Splicing all these exact sequences together, we recover the classical homology long exact sequence associated with the short exact $0 \to C' \to C \to C'' \to 0$.

Theorem 4.16 has a cohomological analogue, which we now state explicitly for the convenience of the reader. A cochain complex C is filtered when it is equipped with a family of subcomplexes $\cdots \subset F^pC \subset F^{p-1}C \subset \cdots$ (beware of the indices, which are decreasing for filtrations of cohomological complexes). The filtration is bounded below if for each k there is an integer p_k such that $F^{p_k}C^k = 0$, and it is *exhaustive* if $\bigcup_{p\in\mathbb{Z}} F^pC = C$. The cohomology of a filtered cochain complex C is filtered by

$$F^pH^*(C) := \mathrm{Im}\big(H^*(F^pC) \to H^*(C)\big)\,.$$

Theorem 4.18. *Let C be a filtered cochain complex, whose filtration is bounded below and exhaustive. There is a cohomological spectral sequence*

$$E_1^{p,q}(C) = H^{p+q}(F^pC/F^{p+1}C) \Longrightarrow H^{p+q}(C)\,.$$

To be more explicit, the differentials are maps $d_r : E_r^{p,q} \to E_r^{p+r,q+1-r}$ and there are isomorphisms:

$$E_\infty^{p,q}(C) \simeq F^p H^{p+q}(C)/F^{p+1}H^{p+q}(C)\,.$$

If $f : C \to C'$ preserves the filtrations, then f induces a morphism of spectral sequences with $E_1^{p,q}(f) = H^{p+q}(\mathrm{Gr}^p f)$, and $E_\infty^{p,q}(f)$ coincides on the abutment with $\mathrm{Gr}^p H^{p+q}(f)$.

The following instructive exercise is suggested by W. van der Kallen. It provides a conceptual interpretation of the terms of the rth page of the spectral sequence of a filtered complex.

Exercise 4.19. Let $f : C \to D$ be a map of cochain complexes. If $f^i : C^i \to D^i$ is surjective and $f^{i+1} : C^{i+1} \to D^{i+1}$ is injective, show that $H^i(f)$ is surjective and $H^{i+1}(f)$ is injective.

If C is a filtered complex, for the sake of brevity, we let $C_{\geq i} = F^i C$, and $C_{i/j} = F^i C/F^j C$ if $j > i$. Let a, b be integers with $a < b$. We have a map ϕ of spectral sequences from the spectral sequence of the filtered complex C to the spectral sequence of the filtered complex $C/C_{\geq b}$, filtered by the images of the $C_{\geq i}$. Show that ϕ_r^{pq} is surjective for $p + r \leq b$ and injective for $p < b$.

Similarly we have a map ψ of spectral sequences from the spectral sequence of the suitably filtered complex $C_{\geq a}/C_{\geq b}$ to the spectral sequence of the filtered complex $C/C_{\geq b}$. Show that ψ_r^{pq} is surjective for $p \geq a$ and injective for $p - r + 1 \geq a$.

Show that the spectral sequence for $C_{\geq a}/C_{\geq b}$ has vanishing E_r^{pq} for $p < a$ and for $p \geq b$.

Now prove the formula $E_r^{pq} = (H^{p+q}(C_{p-r+1/p+r}))_{p/p+1}$ by taking $a = p - r + 1$, $b = p + r$.

4.5.2. Bicomplexes. A *first quadrant chain bicomplex* is a bigraded object $C = \bigoplus_{p\geq 0, q\geq 0} C_{p,q}$, together with differentials $d : C_{p,q} \to C_{p-1,q}$ and $\partial : C_{p,q} \to C_{p,q-1}$ which commute: $d \circ \partial = \partial \circ d$. The *total complex* associated such a bicomplex is the complex $\mathrm{Tot}\, C$ with $(\mathrm{Tot}\, C)_n = \bigoplus_{p+q=n} C_{p,q}$ and whose differential maps an element of $x \in C_{p,q}$ to $dx + (-1)^p \partial x$.

Example 4.20. Let C and D be nonnegative chain complexes of R-modules. Their tensor product $C \otimes_R D$ is equal to $\mathrm{Tot}\,(C \boxtimes_R D)$, where $C \boxtimes_R D$ is the first quadrant chain bicomplex defined by $(C \boxtimes_R D)_{p,q} = C_p \otimes D_q$, $d = d_C \otimes \mathrm{Id}$ and $\partial = \mathrm{Id} \otimes d_D$.

If C is a first quadrant bicomplex, we can obtain a spectral sequence converging to the homology of $\mathrm{Tot}\, C$ as a particular case of Theorem 4.16. Let us sketch the construction. We first define a filtration of $\mathrm{Tot}\, C$ by

$$F_n \mathrm{Tot}\, C := \bigoplus_{0 \leq p,\, 0 \leq q \leq n} C_{p,q}\,.$$

Thus $F_0 C$ equals the 0th column of the bicomplex C, with differential ∂, and more generally F_n is the 'totalization' of the sub-bicomplex of C formed by the objects of columns 0 to n. So there is an isomorphism of complexes:

$$F_n\mathrm{Tot}\,(C)/F_{n-1}\mathrm{Tot}\,(C) \simeq (C_{n,*-n}, \partial)\ .$$

By Theorem 4.16, the filtered complex $\mathrm{Tot}\,C$ gives birth to a spectral sequence starting with page $E^1_{p,q}(C) = H_q(C_{p,*}, \partial)$, converging to $\mathrm{Tot}\,(C)_{p+q}$. Looking at the construction of the spectral sequence of Theorem 4.16, one finds that the first differential $d^1 : E^1_{p,q}(C) \to E^1_{p-1,q}(C)$ is equal to the map $H_q(C_{p,*}, \partial) \to H_q(C_{p-1,*}, \partial)$ induced by the differential $d : C_{p,q} \to C_{p-1,q}$. Thus the second page is given by:

$$E^2_{p,q}(C) = H_p(H_q(C, \partial), d)\ .$$

This is summarized in the following statement.

Theorem 4.21. *Let (C, d, ∂) be a first quadrant chain bicomplex. There is a homological spectral sequence*

$$E^2_{p,q}(C) = H_p(H_q(C, \partial), d) \Longrightarrow H_{p+q}(\mathrm{Tot}\,C)\ .$$

To be more specific, we have isomorphisms:

$$E^\infty_{p,q}(C) \simeq F_pH_{p+q}(\mathrm{Tot}\,C)/F_{p-1}H_{p+q}(\mathrm{Tot}\,C)\ .$$

Moreover, the spectral sequence is natural with respect to morphisms of bicomplexes.

First quadrant chain bicomplexes are 'symmetric', that is, the transposed bicomplex C^t defined by $C^t_{p,q} = C_{q,p}$ gives rise to another spectral sequence with second page $E^2_{p,q}(C^t) = H_q(H_p(C, d), \partial)$ and which converges to $H_{p+q}(C^t) \simeq H_{p+q}(C)$. So there are actually two spectral sequences associated to a bicomplex.

Example 4.22. As an application of Theorem 4.21 we prove that $\mathrm{Tor}^R_*(M, N)$ is 'well balanced', i.e., it can be computed using indifferently a projective resolution of M or a projective resolution of N. Let P (resp. Q) be a projective resolution of M (resp. N). We denote by ${}^\ell\mathrm{Tor}^R_*(M, N)$ the homology of the complex $P \otimes_R N$, and by ${}^r\mathrm{Tor}^R_*(M, N)$ the homology of $M \otimes_R Q$. We form the bicomplex

$$(C, d, \partial) = (P \boxtimes_R Q, d_P \otimes \mathrm{Id}, \mathrm{Id} \otimes d_Q)\ .$$

Theorem 4.21 yields a spectral sequence $(E^r_{p,q}(C), d_r)_{r\geq 2}$, converging to the homology of $\mathrm{Tot}\,(C)$. Since the objects of P are projectives $P_p\otimes_R$ is an exact functor so $H_q(P_p \boxtimes_R Q, \partial) = P_p \otimes_R H_q(Q) = P_p \otimes N$ if $q = 0$ and zero otherwise. Thus:

$$E^2_{p,q}(C) = {}^\ell\mathrm{Tor}_p(M, N) \text{ if } q = 0 \text{ and zero if } q \neq 0.$$

The shape of the second page implies that the differentials d^r, for $r \geq 2$ must be zero (their source or their target is zero), so $E^2_{p,q}(C) = E^\infty_{p,q}(C)$ and since the spectral sequence abuts to $H_*(\mathrm{Tot}\,C)$, we obtain an isomorphism

$${}^\ell\mathrm{Tor}^R_*(M, N) \simeq H_*(\mathrm{Tot}\,C)\ .$$

The same argument applied to the transposed bicomplex C^t yields an isomorphism:

$${}^r\mathrm{Tor}^R_*(M, N) \simeq H_*(\mathrm{Tot}\,C^t) \simeq H_*(\mathrm{Tot}\,C)\ .$$

This proves that $\mathrm{Tor}_*^R(M,N)$ is well balanced. (If $\mathcal{C}$ is a small category, the same reasoning also shows that $\mathrm{Tor}_*^{\mathcal{C}}(G,F)$ is well balanced.)

Exercise 4.23.

1. Let R be an algebra over a commutative ring $\Bbbk$. Let M be a right R-module and let P be a complex of projective left R-modules. Show that there is a first quadrant homological spectral sequence of $\Bbbk$-modules
$$E^2_{p,q} = \mathrm{Tor}_p(M, H_q(P)) \Longrightarrow H_{p+q}(M \otimes_R P)\,.$$
(Hint: consider the bicomplex $Q^M \otimes_R P$, where Q^M is a projective resolution of M.)
2. Assume that M has a projective resolution of length 2 (i.e., of the form $0 \to Q_1 \to Q_0$). Show that if P is a complex of projective R-modules, there are short exact sequences of $\Bbbk$-modules:
$$0 \to M \otimes_R H_q(P) \to H_q(M \otimes_R P) \to \mathrm{Tor}_1^R(M, H_{q-1}(P)) \to 0\,.$$

Theorem 4.21 has the following cohomological analogue for first quadrant cochain bicomplexes (i.e., bigraded objects $C = \bigoplus_{p\geq 0, q\geq 0} C^{p,q}$ equipped with differentials $d : C^{p,q} \to C^{p+1,q}$ and $\partial : C^{p,q} \to C^{p,q+1}$ which commute).

Theorem 4.24. *Let (C, d, ∂) be a first quadrant cochain bicomplex. There is a cohomological spectral sequence*
$$E_2^{p,q}(C) = H^p(H^q(C,\partial), d) \Longrightarrow H^{p+q}(\mathrm{Tot}\, C)\,.$$
To be more specific, the rth differential is a map $E_r^{p,q} \to E_r^{p+r,q+1-r}$ and we have isomorphisms:
$$E_\infty^{p,q}(C) \simeq F^p H^{p+q}(\mathrm{Tot}\, C)/F^{p+1}H^{p+q}(\mathrm{Tot}\, C)\,.$$
Moreover, the spectral sequence is natural with respect to morphisms of bicomplexes.

4.5.3. Grothendieck spectral sequences. Let $\mathcal{A}$, $\mathcal{B}$, $\mathcal{C}$ be abelian categories, and let $F : \mathcal{A} \to \mathcal{B}$ and $G : \mathcal{B} \to \mathcal{C}$ be right exact functors. On can wonder wether it is possible to recover the derived functors $L_i(G \circ F)$ of the composition from the derived functors L_iF and L_jG. The Grothendieck spectral sequence answers this question under an acyclicity assumption. One says that $N \in \mathcal{B}$ is *G-acyclic* if $L_iG(N) = 0$ for $i > 0$.

Theorem 4.25. *Let $\mathcal{A}$, $\mathcal{B}$, $\mathcal{C}$ be abelian categories such that $\mathcal{A}$ and $\mathcal{B}$ have enough projectives. Assume that we have right exact functors $F : \mathcal{A} \to \mathcal{B}$ and $G : \mathcal{B} \to \mathcal{C}$ such that F sends projective objects to G-acyclic objects. Then for all $M \in \mathcal{A}$, there is a homological spectral sequence:*
$$E^2_{p,q} = L_pG \circ L_qF(M) \Longrightarrow L_{p+q}(G \circ F)(M)\,.$$

The Grothendieck spectral sequence has a cohomological analogue for right derived functors.

Theorem 4.26. *Let $\mathcal{A}$, $\mathcal{B}$, $\mathcal{C}$ be abelian categories such that $\mathcal{A}$ and $\mathcal{B}$ have enough injectives. Assume that we have left exact functors $F : \mathcal{A} \to \mathcal{B}$ and $G : \mathcal{B} \to \mathcal{C}$ such that F sends injective objects to G-acyclic objects. Then for all $M \in \mathcal{A}$, there is a cohomological spectral sequence, with differentials $d_r : E_r^{p,q} \to E_r^{p+r,q+1-r}$:*

$$E_2^{p,q} = R^pG \circ R^qF(M) \Longrightarrow R^{p+q}(G \circ F)(M)\ .$$

Example 4.27. Let us illustrate the Grothendieck spectral sequence by a situation from the algebraic group setting. Let G be an algebraic group over a field $\Bbbk$ and let H be a closed subgroup. The restriction functor:

$$\mathrm{res}_H^G : \{\text{rat. } G\text{-mod}\} \to \{\text{rat. } H\text{-mod}\}$$

admits a right adjoint, namely the induction functor[10]

$$\mathrm{ind}_H^G : \{\text{rat. } H\text{-mod}\} \to \{\text{rat. } G\text{-mod}\}\ .$$

By Exercise 2.22, ind_H^G is left exact and preserves the injectives. Moreover, $H^*(H, M)$ is the derived functor of

$$M \mapsto \mathrm{Hom}_{\text{rat. } H\text{-mod}}(\Bbbk, M) \simeq \mathrm{Hom}_{\text{rat. } G\text{-mod}}(\Bbbk, \mathrm{ind}_H^G M)$$

so Theorem 4.26 yields a spectral sequence:

$$E_2^{p,q} = H^p(G, R^q\mathrm{ind}_H^G(M)) \Longrightarrow H^{p+q}(H, M)\ .$$

In particular, if by chance the H-module M satisfies the vanishing condition $R^q\mathrm{ind}_H^G(M) = 0$ for $q > 0$, the spectral sequence stops at the second page (since it is concentrated on the 0th row) and yields an isomorphism:

$$H^*(G, \mathrm{ind}_H^G M) \simeq H^*(H, M)\ .$$

The Kempf vanishing theorem, a cornerstone for the representation theory of algebraic groups, asserts that the vanishing condition is satisfied when G is reductive, H is a Borel subgroup of G and $M = \Bbbk_\lambda$ is the one-dimensional representation of H given by a character χ_λ of H, associated to a dominant weight λ[11]. In this case, the representation $\mathrm{ind}_H^G(\Bbbk_\lambda)$ is called 'the costandard module with highest weight λ' and denoted by $\nabla_G(\lambda)$, or $H^0(\lambda)$ [J, II.2].

---

[10] Let us consider the regular functions $\Bbbk[G]$ as a $G \times H$-module, where G acts by left translations and H by right translations on G. The induction functor is defined by the formula: $\mathrm{ind}_H^G(M) = (\Bbbk[G] \otimes M)^H$.

[11] For example, $G = GL_n(\Bbbk)$, $H = B_n(\Bbbk)$ is the subgroup of upper triangular matrices and $M = \Bbbk_\lambda$ is the one-dimensional vector space $\Bbbk$ acted on by $B_n(\Bbbk)$ by multiplication with a scalar $\chi_\lambda([g_{i,j}]) = \prod g_{i,i}^{\lambda_i}$, where $\lambda_1 \geq \lambda_2 \geq \cdots \geq \lambda_n$.

4.5.4. Filtered differential graded algebras. A differential graded algebra is a graded algebra A equipped with a differential d satisfying the Leibniz rule: $d(xy) = d(x)y + (-1)^{\deg x} x d(y)$. A filtration of a differential graded algebra A is a filtration $\cdots \subset F_pA \subset F_{p+1}A \subset \cdots A$ of the complex (A, d) compatible with products, i.e., the product sends $F_pA \otimes F_qA$ into $F_{p+q}A$.

If A is a filtered differential algebra, the product of A induces bigraded algebra structures on $\bigoplus_{p,q} H_q(\mathrm{Gr}_pA)$ and $\bigoplus_{p,q} \mathrm{Gr}_pH_q(A)$. Thus, the first page and the ∞-page of the spectral sequence of Theorem 4.16 are bigraded algebras. More is true: in this situation, the spectral sequence is a spectral sequence of algebras.

Theorem 4.28. *Let A be a filtered differential graded algebra, whose filtration is bounded below and exhaustive. There is a homological spectral sequence of algebras*

$$E^1_{p,q}(A) = H_{p+q}(\mathrm{Gr}_pA) \Longrightarrow H_{p+q}(A) .$$

More explicitly, there are bigraded algebra isomorphisms

$$E^\infty_{p,q}(A) \simeq F_pH_{p+q}(A)/F_{p-1}H_{p+q}(A) .$$

If $f : A \to A'$ is a morphism of differential graded algebras preserving the filtrations (i.e., it sends F_pA into F_pA'), then f induces a morphism of spectral sequences with $E^1_{p,q}(f) = H_{p+q}(\mathrm{Gr}_pf)$, and $E^\infty_{p,q}(f)$ coincides on the abutment with $\mathrm{Gr}_pH_{p+q}(f)$.

Theorem 4.28 has an obvious cohomological analogue, whose formulation is left to the reader.

4.5.5. The Lyndon–Hochschild–Serre spectral sequence. Let G be a discrete group, let H be a normal subgroup of G, and let M be a $\Bbbk G$-module ($\Bbbk$ is a commutative ring). The Lyndon–Hochschild–Serre spectral sequence allows to reconstruct the cohomology $H^*(G, M)$ from some cohomology groups of H and G/H. Let us be more specific. We can restrict the action on M to G to obtain a $\Bbbk H$-module, still denoted by M.

Lemma 4.29. *The quotient group G/H acts on $H^i(H, M)$ for all $i \geq 0$.*

Proof. Let $[g]$ denote the class of $g \in G$ in G/H. Then G/H acts on M^H by the formula: $[g]m := gm$. Consider the cohomological δ-functor $(F^i, \delta^i)_{i\geq 0}$, where the F^i are the functors

$$\begin{array}{ccc} \Bbbk G\text{-Mod} & \to & \Bbbk\text{-Mod} \\ M & \mapsto & H^i(H, M) \end{array} .$$

Restriction from $\Bbbk G$-modules to $\Bbbk H$-modules preserves injectives (cf. Exercise 2.31), so $F^i = R^i(F^0)$. Each $[g] \in G/H$ defines a natural transformation $[g] : F^0 \to F^0$ so it extends uniquely into a morphism of δ-functors from $(F^i, \delta^i)_{i\geq 0}$ to $(F^i, \delta^i)_{i\geq 0}$. By uniqueness, since the axioms of a group action are satisfied on $F^0(M)$, they are satisfied on all $F^i(M)$, $i \geq 0$. □

We are now ready to describe the Lyndon–Hochschild–Serre spectral sequence.

Theorem 4.30. *Let H be a normal subgroup of G, and let M be a $\Bbbk G$-module. There is a first quadrant cohomological spectral sequence of $\Bbbk$-modules, with differentials $d^r : E_r^{p,q} \to E_r^{p+r,q+r-1}$:*

$$E_2^{p,q} = H^p(G/H, H^q(H,M)) \Longrightarrow H^{p+q}(G,M)\ .$$

This spectral sequence is natural with respect to M. Moreover, if M is a $\Bbbk G$-algebra then the spectral sequence is a spectral sequence of algebras. (On the second page of the spectral sequence, the product is the cup product of the cohomology algebra $H^*(G/H, A)$ where A is the cohomology algebra $H^*(H,M)$.)

A similar Lyndon–Hochschild–Serre spectral sequence exists for the homology of discrete groups, and the rational cohomology of algebraic groups.

Exercise 4.31. Let H be a normal subgroup of G and let $M \in \Bbbk G$-mod ($\Bbbk$ is a commutative ring). Derive from the Lyndon–Hochschild–Serre spectral sequence the five terms exact sequence:

$$\begin{aligned} 0 \to H^1(G/H, M^H) \to H^1(G,M) \to H^1(H,M)^{G/H} \\ \to H^2(G/H, M^H) \to H^2(G,M). \end{aligned}$$

References

[Ben] D. Benson, Representations and cohomology. II. Cohomology of groups and modules. Second edition. Cambridge Studies in Advanced Mathematics, 31. Cambridge University Press, Cambridge, 1998.

[Boa] M. Boardman, Conditionally convergent spectral sequences. Homotopy invariant algebraic structures (Baltimore, MD, 1998), 49–84, Contemp. Math., 239, Amer. Math. Soc., Providence, RI, 1999.

[Bo] A. Borel, Linear algebraic groups. Second edition. Graduate Texts in Mathematics, 126. Springer-Verlag, New York, 1991.

[Br] K. Brown, Cohomology of groups. Graduate Texts in Mathematics, 87. Springer-Verlag, New York-Berlin, 1982.

[DP] A. Dold, D. Puppe, Homologie nicht-additiver Funktoren; Anwendugen. Ann. Inst. Fourier 11, 1961, 201–312.

[Do] A. Dold, Homology of symmetric products and other functors of complexes. Ann. of Math. (2) 68 1958, 54–80.

[F] E. Friedlander, Lectures on the cohomology of finite group schemes. Rational representations, the Steenrod algebra and functor homology, 27–53, Panor. Synthèses, 16, Soc. Math. France, Paris, 2003.

[FP] V. Franjou, T. Pirashvili, Stable K-theory is bifunctor homology (after A. Scorischenko). Rational representations, the Steenrod algebra and functor homology, 27–53, Panor. Synthèses, 16, Soc. Math. France, Paris, 2003.

[GJ] P. Goerss, J. Jardine, Simplicial Homotopy Theory. Progress in Mathematics, Birkhäuser, Basel-Boston-Berlin, 1999.

[Hat] A. Hatcher, Algebraic Topology. Cambridge, England: Cambridge University Press, 2002.

[Hov] M. Hovey, Model categories. Mathematical Surveys and Monographs, 63. American Mathematical Society, Providence, RI, 1999.

[Hum] J. Humphreys, Linear algebraic groups. Graduate Texts in Mathematics, No. 21. Springer-Verlag, New York-Heidelberg, 1975.

[J] J.C. Jantzen, Representations of algebraic groups. Second edition. Mathematical Surveys and Monographs, 107. American Mathematical Society, Providence, RI, 2003.

[May] P. May, Simplicial objects in algebraic topology. Princeton: Van Nostrand, 1967.

[Mit] B. Mitchell, Rings with several objects. Advances in Math. 8 (1972), 1–161.

[MC] J. McCleary, A user's guide to spectral sequences. Second edition. Cambridge Studies in Advanced Mathematics, 58. Cambridge University Press, Cambridge, 2001.

[ML] S. Mac Lane, Homology. Reprint of the 1975 edition. Classics in Mathematics. Springer-Verlag, Berlin, 1995.

[Q] D. Quillen, Homotopical algebra. Lecture Notes in Mathematics, No. 43 Springer-Verlag, Berlin-New York 1967.

[S] T.A. Springer, Linear algebraic groups. Reprint of the 1998 second edition. Modern Birkhäuser Classics. Birkhäuser Boston, Inc., Boston, MA, 2009.

[Wa] W.C. Waterhouse, Introduction to affine group schemes. Graduate Texts in Math. 66, Springer-Verlag, Berlin-New York, 1994.

[Wei] C. Weibel, An introduction to homological algebra. Cambridge Studies in Advanced Mathematics, 38. Cambridge University Press, Cambridge, 1994.

Antoine Touzé
Université Lille 1 – Sciences et Technologies
Laboratoire Painlevé
Cité Scientifique – Bâtiment M2
F-59655 Villeneuve d'Ascq Cedex, France
e-mail: `antoine.touze@math.univ-lille1.fr`

Printed in the United States
By Bookmasters